Modern Food Microbiology

Modern Food Microbiology

Ralph Hawkins

MURPHY & MOORE
www.murphy-moorepublishing.com

Published by Murphy & Moore Publishing,
1 Rockefeller Plaza,
New York City, NY 10020, USA
www.murphy-moorepublishing.com

Modern Food Microbiology
Ralph Hawkins

International Standard Book Number: 978-1-63987-372-2 (Hardback)

Cataloging-in-Publication Data

Modern food microbiology / Ralph Hawkins.
p. cm.
Includes bibliographical references and index.
ISBN 978-1-63987-372-2
1. Food--Microbiology. 2. Food--Bacteriology. 3. Sanitary microbiology.
4. Microorganisms. 5. Biology. I. Hawkins, Ralph.
QR115 .M63 2022
664.001 579--dc23

TABLE OF CONTENTS

PREFACE

The branch of biology which deals with the microorganisms responsible for inhibiting, creating and contaminating food is called food microbiology. It focuses on the microbes and pathogens responsible for food spoilage, diseases, and production of fermented foods like cheese, beer, yoghurt, bread and wine. Depending upon their application, bacteria studied under this discipline can be classified into lactic acid bacteria, acetic acid bacteria, proteolytic bacteria, lipolytic bacteria, saccharolytic bacteria, etc. The major areas of research under food microbiology are food safety, food testing and fermentation. The development of biopolymers such as alginate and poly-γ-glutamic acid is also undertaken within this discipline. This book presents the complex subject of food microbiology in the most comprehensible and easy to understand language. Some of the diverse topics covered herein address the varied branches that fall under this category. The readers would gain knowledge that would broaden their perspective about modern food microbiology.

A detailed account of the significant topics covered in this book is provided below:

Chapter 1- The branch of microbiology which is concerned with the study of micro-organisms in food and for the production of food is known as food microbiology. A few of the important aspects associated with this domain are micro-organisms in food and microbial growth. This is an introductory chapter which will briefly introduce all these significant aspects of food microbiology.

Chapter 2- The process which makes the food undesirable and unacceptable for consumption is known as food spoilage. Some of the important topics studied under food spoilage are meat and poultry spoilage, seafood spoilage, fruit and vegetable spoilage, and milk and dairy products' spoilage. This chapter has been carefully written to provide an easy understanding of these varied facets of food spoilage.

Chapter 3- Fermentation refers to the metabolic process of breakdown of an organic substrate by the micro-organisms through the action of enzymes. Fermented foods can be divided into animal-based fermented foods, plant-based fermented foods and fermented drinks. The topics elaborated in this chapter will help in gaining a better perspective about fermentation and fermented foods.

Chapter 4- All the techniques of food processing that prevent the spoilage of food due to the growth of micro-organisms and slow down the oxidation of fats fall under food preservation. It includes frozen foods, pickling, food sampling and home canning. The chapter closely examines these key concepts of food preservation to provide an extensive understanding of the subject.

Chapter 5- Food storage refers to the processes which help in storing food in such a manner that it can be utilized at a later stage for consumption. This chapter closely examines the key concepts of food storage to provide an extensive understanding of the subject. Some of the topics elaborated in this chapter are food bank, food storage containers, sweet potato storage, etc.

Chapter 6- Any disease caused by food contaminated with pathogenic bacteria, viruses or parasites is known as microbial foodborne disease. Some of the significant topics studied under microbial foodborne diseases are foodborne illness, foodborne infection and intoxication, prevention of foodborne illness, treatment of foodborne illness, etc. These topics are crucial for a complete understanding of the subject.

I would like to make a special mention of my publisher who considered me worthy of this opportunity and also supported me throughout the process. I would also like to thank the editing team at the back-end who extended their help whenever required.

Ralph Hawkins

Chapter 1

Food Microbiology: An Introduction

The branch of microbiology which is concerned with the study of micro-organisms in food and for the production of food is known as food microbiology. A few of the important aspects associated with this domain are micro-organisms in food and microbial growth. This is an introductory chapter which will briefly introduce all these significant aspects of food microbiology.

Except for a few sterile foods, all foods harbor one or more types of microorganisms. Some of them have desirable roles in food, such as in the production of naturally fermented food, whereas others cause food spoilage and foodborne diseases. To study the role of microorganisms in food and to control them when necessary, it is important to isolate them in pure culture and study their morphological, physiological, biochemical, and genetic characteristics. Some of the simplest techniques in use today for these studies were developed over the last 300 years.

Discovery of Microorganisms

The discovery of microorganisms ran parallel with the invention and improvement of the microscope. Athanasius Kircher reported that, using a microscope, he had seen minute living worms in putrid meat and milk. The magnification power of his microscope was so low that he could not have seen bacteria. In 1664, Robert Hooke described the structure of molds. However, probably the first person to see different types of microorganisms, especially bacteria, under a microscope that probably did not have a magnification power above 300v, was Antony van Leeuwenhoek. He observed bacteria in saliva, rainwater, vinegar, and other materials; sketched the three morphological groups (spheroids or cocci, cylindrical rods or bacilli, and spiral or spirilla) and also described some to be motile. He called them animalcules.

As reasonably good microscopes were not easily available at the time, other interested individuals and scientists during the next 100 years only confirmed Leeuwenhoek's observations. In the 19th century, as an outcome of the Industrial Revolution, improved microscopes became more easily available, which stimulated many inquisitive minds to observe and describe the creatures they discovered under a microscope. By 1838, Ehrenberg (who introduced the term bacteria) had proposed at least 16 species in four genera and by 1875 Ferdinand Cohn had developed the preliminary classification system of bacteria. Cohn also was the first to discover that some bacteria produced spores. Although, like bacteria, the existence of sub-microscopic viruses was recognized in the mid19th century, they were observed only after the invention of the electron microscope in the 1940s.

Where are they Coming from?

Following Leeuwenhoek's discovery, although there were no bursts of activity, some scientific minds were curious to determine from where the animalcules, observed to be present in many different objects, were emanating. Society had just emerged from the Renaissance period, and science, known as experimental philosophy, was in its infancy. The theory of spontaneous generation, i.e., the generation of some form of life from non-living objects, had many powerful followers among the educated and elite classes. Since the time of the Greeks, the emergence of maggots from dead bodies and spoiled flesh was thought to be due to spontaneous generation. However, Redi disproved that theory by showing that the maggots in spoiled meat and fish could only appear if flies were allowed to contaminate them. The advocates of the spontaneous generation theory argued that the animalcules could not regenerate by themselves (biogenesis), but they were present in different things only through abiogenesis (spontaneous generation). In 1749, Turbevill Needham showed that boiled meat and meat broth, following storage in covered flasks, could have the presence of animalcules within a short time. This was used to prove the appearance of these animalcules by spontaneous generation. Lazzaro Spallanzani showed that boiling the meat infusion in broth in a flask and sealing the flask immediately prevented the appearance of these microscopic organisms, thereby disproving Needham's theory.

This was the time when Antoine Laurent Lavoisier and his co-workers showed the need of oxygen for life. The believers of abiogenesis rejected Spallanzani's observation, suggesting that there were not enough vital forces (oxygen) present in the sealed flask for animalcules to appear through spontaneous generation. Later, Schulze, Theodore Schwann and Schröeder showed that bacteria failed to appear in boiled meat infusion even in the presence of air. Finally, in 1861, Louis Pasteur demonstrated that, in boiled infusion, bacteria could grow only if the infusions were contaminated with bacteria carried by dust particles in air. His careful and controlled studies proved that bacteria were able to reproduce (biogenesis) and life could not originate by spontaneous generation. John Tyndall, in 1870, showed that boiled infusion could be stored in dust-free air in a box without microbial growth.

What are their Functions?

The involvement of invisible organisms in many diseases in humans was suspected as early as the 13th century by Roger Bacon. In the 16th century, Girolamo Fracastoro of Verona suggested that many human diseases were transmitted from person to person by small creatures. This was also indicated by Kircher in 1658. In 1762, von Plenciz of Vienna suggested that different invisible organisms were responsible for different diseases. Theodore Schwann and Hermann Helmholtz proposed that putrefaction and fermentation were connected with the presence of the organisms derived from air. Finally, Pasteur, in 1875, showed that wine fermentation from grapes and souring of wine were caused by microorganisms. He also proved that spoilage of meat and milk was associated with the growth of microorganisms. Later, he showed the association of microorganisms with several diseases in humans, cattle, and sheep and he also developed vaccines against a few human and animal diseases caused by microorganisms, including rabies. Robert Koch, in Germany, isolated pure cultures of bacteria responsible for anthrax, cholera, and tuberculosis. He also developed the famous Koch's postulates to associate a specific bacterium as a causative agent for a specific disease. Along with his associates, he also developed techniques of agar plating methods to isolate bacteria in pure cultures and to determine microbial numbers in a sample, the Petri dish, staining methods for better microscopic observation of bacteria, and the use of steam to sterilize materials to grow bacteria.

With time, the importance of microorganisms in human and animal diseases, soil fertility, plant diseases, fermentation, food spoilage and foodborne diseases, and other areas was recognized and microbiology was developed as a specific discipline. Later, it was divided into several sub-disciplines, such as medical microbiology, soil microbiology, plant pathology, and food microbiology.

Development of Early Food Microbiology (Before 1900 A.D.)

It is logical to comprehend that our early Homo ancestors, the hunters and gatherers, were aware of food spoilage and foodborne diseases. Even without any perception of the causative agents, they used ice and fire to preserve foods and make them safe. Around 8000 B.C., as agriculture and animal husbandry were adopted by the early civilizations, food supply, especially agricultural produce, became available in abundance during the growing seasons. Preservation of foods became important for uniform supply of food around the year. Between 8000 and 1000 B.C. many food preservation methods such as drying, cooking, baking, smoking, salting, sugaring (with honey), low-temperature storage (in ice), storage without air (in pits), fermentation (with fruits, grains, and milk), pickling, and spicing were used, probably mainly to reduce spoilage.

However, one cannot be sure whether the society at that time recognized the implications of diseases transmitted through food. In the later periods, however, the scriptural injunctions laid by many religions suggest that the societies recognized an association of diseases with some foods. Some of the regulations, such as not eating meat from a diseased animal or an animal killed by a scavenger, or not eating a food that appeared unnatural or had been handled by an unclean person, were developed to safeguard the health of citizens against foodborne diseases. Fermentation was used extensively by many societies not only to preserve foods but also as a method to produce various types of desirable foods from milk, meat, fish, eggs, grains, fruits, and vegetables.

Following the discovery of the ubiquitous existence of microorganisms (mainly bacteria and yeasts) by Leeuwenhoek around the 1670s, some individuals started associating the possible role of these organisms with food spoilage, food fermentation, and foodborne diseases. The major developments of ideas on the possible roles of microorganisms in foods and their scientific proof were initiated by Pasteur in the 1870s, followed by many other scientists before the end of the 19th century. This paved the way for the establishment of early food microbiology in the 20th century. Some of the major developments in the 19th century are briefly listed here:

Food Fermentation	
1822	C.J. Person named the microscopic organism found on the surface of wine during vinegar production as Mycoderma mesentericum. Pasteur in 1868 proved that this organism was associated with the conversion of alcohol to acetic acid and named it Mycoderma aceti. In 1898, Martinus Beijerinck renamed it Acetobacter aceti.

1837	Theodor Schwann named the organism involved in sugar fermentation as Saccharomyces (sugar fungus).
1838	Charles Cogniard-Latour suggested that growth of yeasts was associated with alcohol fermentation.
1860	Louis Pasteur showed that fermentation of lactic acid and alcohol from sugar was the result of growth of specific bacteria and yeasts, respectively.
1883	Emil Christian Hansen used pure cultures of yeasts to ferment beer.

Food Spoilage	
1804	Francois Nicolas Appert developed methods to preserve foods in sealed glass bottles by heat in boiling water. He credited this process to Lazzaro Spallanzani, who first used the method to disprove the spontaneous generation theory.
1819	Peter Durand developed canning preservation of foods in steel cans. Charles Mitchell introduced tin lining of metal cans in 1839.
1870	Peter Durand developed canning preservation of foods in steel cans. Charles Mitchell introduced tin lining of metal cans in 1839.
1895	Harry Russell showed that gaseous swelling with bad odors in canned peas was due to growth of heat-resistant bacteria (spores).

Foodborne Diseases	
1820	Justin Kerner described food poisoning from eating blood sausage (due to botulism). Fatal disease from eating blood sausage was recognized as early as A.D. 900.
1849	John Snow suggested the spread of cholera through drinking water contaminated with sewage. In 1854, Filippo Facini named the cholera bacilli as Vibrio cholera, which was isolated in pure form by Robert Koch in 1884.
1856	William Budd suggested that water contamination with feces from infected person spread typhoid fever and advocated the use of chlorine in water supply to overcome the problem. In 1800, G. de Morveau and W. Cruikshank advocated the use of chlorine to sanitize potable water.
1885	Theodor Escherich isolated Bacterium coli (later named Escherichia coli) from the feces and suggested that some strains were associated with infant diarrhea.
1888	A.A. Gartner isolated Bacterium (later Salmonella) enteritidis from the organs of a diseased man as well as from the meat the man ate. In 1896, Marie von Ermengem proved that Salmonella enteritidis caused a fatal disease in humans who consumed contaminated sausage.
1894	J. Denys associated pyogenic Staphylococcus with death of a person who ate meat prepared from a diseased cow.
1895	Marie von Ermengem isolated Bacillus botulinus (Clostridium botulinum) from contaminated meat and proved that it caused botulism.

Microbiology Techniques	
1854	Heinrich Schröder and Theodore von Dusch used cotton to close tubes and flasks to prevent microbial contamination in heated culture broths.
1876	Car Weigert used methylene blue (a synthetic dye) to stain bacteria in aqueous suspensions.
1877	Ferdinand Cohn showed heat resistance of Bacillus subtilis endospores.
1878	Joseph Lister isolated Streptococcus (now Lactococcus) lactis in pure culture by serial dilution from sour milk.
1880s	Robert Koch and his associates introduced many important methods that are used in all branches of microbiology, such as solid media (first gelatin, then agar) to purify and enumerate bacteria, Petri dish, flagellar staining, steam sterilization of media above 100rC, and photography of cells and spores.
1884	Hans Christian Gram developed Gram staining of bacterial cells.

In the early 20th century, studies continued to understand the association and importance of microorganisms, especially pathogenic bacteria in food. Specific methods were developed for their isolation and identification. The importance of sanitation in the handling of food to reduce contamination by microorganisms was recognized. Specific methods were studied to prevent growth as well as to destroy the spoilage and pathogenic bacteria. There was also some interest to isolate beneficial bacteria associated with food fermentation, especially dairy fermentation, and study their characteristics. However, after the 1950s, food microbiology entered a new era. Availability of basic information on the physiological, biochemical, and biological characteristics of diverse types of food, microbial interactions in food environments and microbial physiology, biochemistry, genetics, and immunology has helped open new frontiers in food microbiology. Among these are:

Food Fermentation/Probiotics

- Development of strains with desirable metabolic activities by genetic transfer among strains.
- Development of bacteriophage-resistant lactic acid bacteria.
- Metabolic engineering of strains for overproduction of desirable metabolites.
- Development of methods to use lactic acid bacteria to deliver immunity proteins.
- Sequencing genomes of important lactic acid bacteria and bacteriophages for better understanding of their characteristics.
- Food bio-preservation with desirable bacteria and their antimicrobial metabolites.
- Understanding of important characteristics of probiotic bacteria and development of desirable strains.
- Effective methods to produce starter cultures for direct use in food processing.

Food Spoilage

- Identification and control of new spoilage bacteria associated with the current changes in food processing and preservation methods.
- Spoilage due to bacterial enzymes of frozen and refrigerated foods with extended shelf life.
- Development of molecular methods (nanotechnology) to identify metabolites of spoilage bacteria and predict potential shelf life of foods.
- Importance of environmental stress on the resistance of spoilage bacteria to antimicrobial preservatives.

Foodborne Diseases

- Methods to detect emerging foodborne pathogenic bacteria from contaminated foods.
- Application of molecular biology techniques (nanotechnology) for rapid detection of pathogenic bacteria in food and environment.
- Effective detection and control methods of foodborne pathogenic viruses.

- Transmission potentials of prion diseases from food animals to humans Importance of environmental stress on the detection and destruction of pathogens.
- Factors associated with the increase in antibiotic-resistant pathogens in food.
- Adherence of foodborne pathogens on food and equipment surfaces.
- Mechanisms of pathogenicity of foodborne pathogens.
- Effective methods for epidemiology study of foodborne diseases.
- Control of pathogenic parasites in food.

Miscellaneous

- Application of hazard analysis of critical control points (HACCP) in food production, processing, and preservation.
- Novel food-processing technologies.
- Microbiology of unprocessed and low-heat-processed ready-to-eat foods.
- Microbial control of foods from farm to table (total quality management).
- Food safety legislation.

Food Microbiology and Food Microbiologists

Before the 1970s, food microbiology was regarded as an applied science mainly involved in the microbiological quality control of food. Since then, the technology used in food production, processing, distribution and retailing and food consumption patterns have changed dramatically. These changes have introduced new problems that can no longer be solved by merely using applied knowledge. Thus, modern-day food microbiology needs to include a great deal of basic science to understand and effectively solve the microbiological problems associated with food. The discipline includes not only microbiological aspects of food spoilage and foodborne diseases and their effective control and bioprocessing of foods but also basic information of microbial ecology, physiology, metabolism, and genetics. This information is helping to develop methods for rapid and effective detection of spoilage and pathogenic bacteria, to develop desirable microbial strains by recombinant DNA technology, to produce fermented foods of better quality, to develop thermos table enzymes in enzyme processing of food and food additives, to develop methods to remove bacteria from food and equipment surfaces, and to combine several control methods for effective control of spoilage and pathogenic microorganisms in food. An individual who has completed courses in food microbiology should gain knowledge in the following areas:

- Determine microbiological quality of foods and food ingredients by using appropriate techniques.
- Determine microbial types involved in spoilage and health hazards and identify the sources.
- Design corrective procedures to control the spoilage and pathogenic microorganisms in food.

- Learn rapid methods to isolate and identify pathogens and spoilage bacteria from food and environment Identify how new technologies adapted in food processing can have specific microbiological problems and design methods to overcome the problem.
- Design effective sanitation procedures to control spoilage and pathogen problems in food-processing facilities.
- Effectively use desirable microorganisms to produce fermented foods.
- Design methods to produce better starter cultures for use in fermented foods and probiotics.
- Know about food regulations (state, federal, and international).
- Understand microbiological problems of imported foods.

To be effective, in addition to the knowledge gained, one has to be able to communicate with different groups of people about the subject (food microbiology and its relation to food science). An individual with good common sense is always in a better position to sense a problem and correct it quickly.

Microorganisms in Food

Living cellular organisms, on the basis of phylogenetic and evolutionary relationships, are grouped into five kingdoms in which bacteria belong to prokaryote (before nucleus), while the eukaryotic (with nucleus) molds and yeasts are grouped under fungi. Viruses are not considered as living cells and are not included in this classification system. For the classification of yeasts, molds, and bacteria, several ranks are used after the kingdom. These are divisions, classes, orders, families, genera (singular, genus), and species. The basic taxonomic group is the species. Several species with similar characteristics form a genus.

A family is made up of several genera, and the same procedure is followed in the hierarchy. Ranks above species, genus, and family are seldom used in food microbiology. Among bacteria, a species is regarded as a collection of strains having many common features. A strain is the descendent of a single colony (single cell). Among the strains in a species, one is assigned as the type strain; it is used as a reference strain while comparing the characteristics of an unknown isolate.

The basic taxonomic group in bacteria, yeasts, and molds is the species, and each species is given a name. The name has two parts (binomial name); the first part is the genus name and the second part is the specific epithet (adjective). Both parts are Latinized; when written, they are italicised (or underlined) with the first letter of the genus written in a capital letter and species name in small letters. For e.g. Bacillus subtilis (genus is Bacillus and species is subtilis).

Micro-Organisms in Food

The Micro-organisms most common to food are bacteria and fungi. The fungi, which are less common than bacteria, consist of two major types of Microorganisms, viz. molds and yeasts. Apart from these, food may contain viruses and other parasites such as protozoans, worms etc.

Bacteria

Bacteria are unicellular Micro-organisms that are approximately one micro meter (10^{-3}mm) in diameter with variations in morphology from short and elongated rods (bacilli), spherical or ovoid forms(cocci), vibrio (comma shaped) and even spiral in shape. Cocci (meaning "berry") are sphere shaped bacteria. Individual bacteria closely combine in various forms according to genera. Some sphere-shaped bacteria occur in clusters similar to a bunch of grapes (i.e. staphylococci). Other bacteria (rod shaped or sphere shaped) are linked together to form chains (i.e. streptococci in case of cocci chain). Certain genera of sphere-shaped bacteria are found together in pairs (diplococci i.e. Pneumococci) or as a group of four (Square or cubical packets formation; i.e. Sarcinia), while other genera appear as an individual bacterium. Other bacteria (in majority) are rod shaped and possess flagella and are motile.

Bacteria produce various pigments which range from shades of yellow to dark pigments such as brown or black. Certain bacteria have pigmentation of intermediate colors such as red, pink, orange, blue, green, or purple. These bacteria cause food discoloration, especially, among foods with unstable color pigments such as meat. Some bacteria also cause discoloration by slime formation.

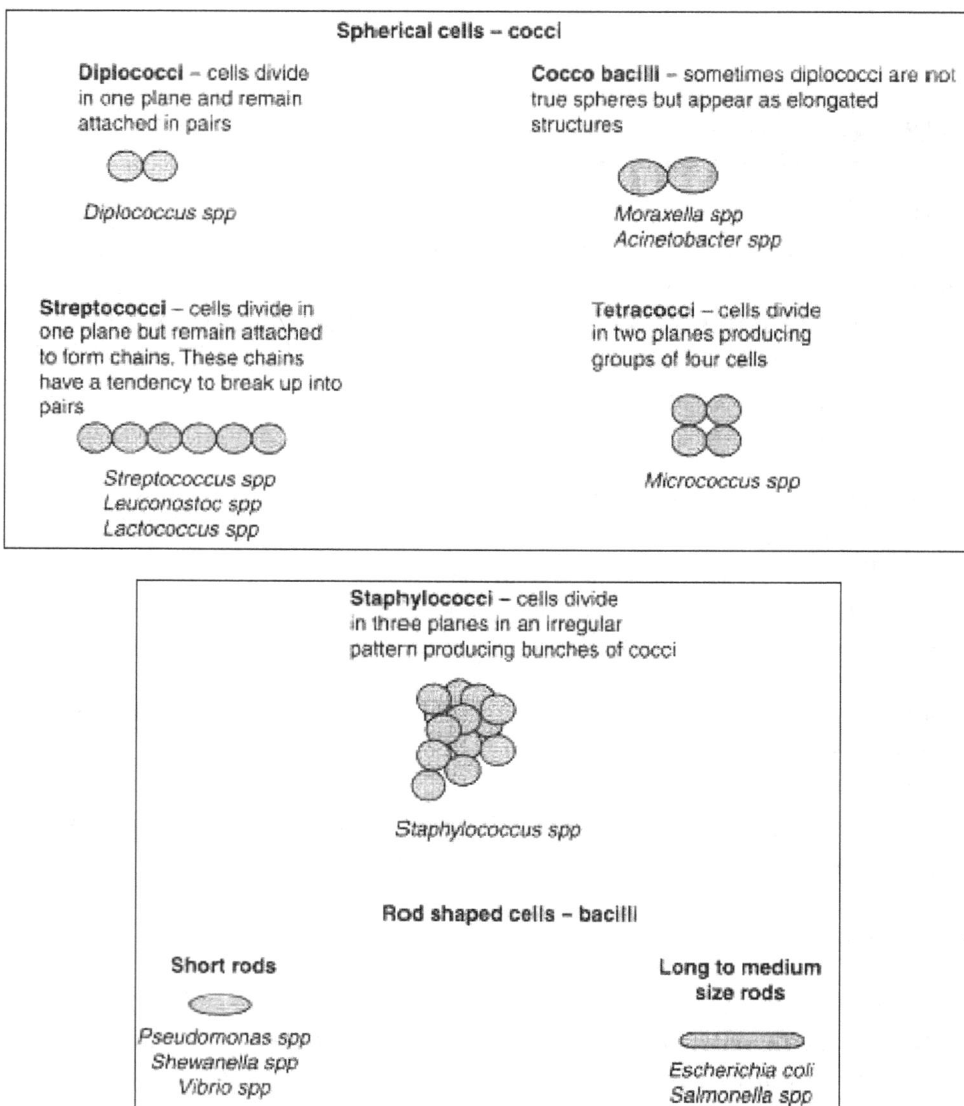

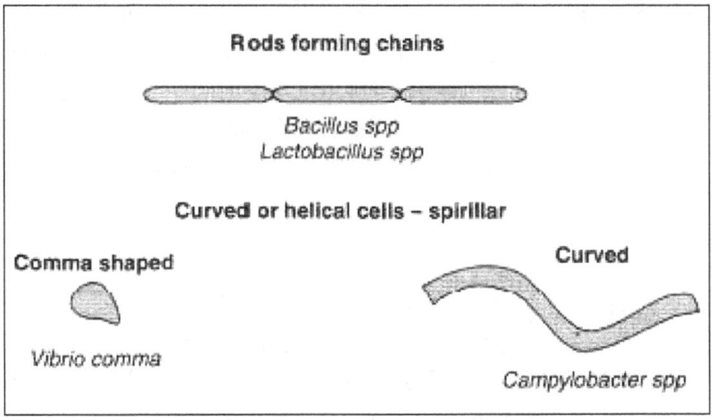

Types of bacterial cells and their groupings.

Molds

Molds are multicellular micro-organisms with mycelial (filamentous) morphology. These microbes are also characterized by their display of a variety of colors and are generally recognized by their mildewy or fuzzy, cotton like appearance. Molds can develop numerous tiny spores that are found in the air and can be spread by air currents. These spores can produce new mold growth if they are transferred to a location that has conditions conducive to germination. Molds generally withstand greater fluctuation in pH than bacteria and yeasts and can frequently tolerate more temperature fluctuation. Although molds thrive best at or near a pH of 7.0, a pH range of 2.0 to 8.0 can be tolerated, even though an acid to neutral pH is preferred. Molds thrive better at ambient temperature than in a colder environment, even though growth can occur below 0 °C. Although mold growth is optimal at a water activity (A_w) of approximately 0.85, growth can and does occur below 0.80. At an A_w of 0.90 or higher, bacteria and yeasts grow more effectively and normally utilize available nutrients for growth at the expense of molds. When the A_w goes below 0.90, molds grow more effectively. That is why foodstuffs, such as pastries, cheeses, and nuts, that are low in moisture content are more likely to spoil from mold growth.

Yeasts

Yeasts are generally unicellular and differ from bacteria in their large cell size and morphology, and because they produce buds during the process of reproduction by division. Like molds, yeasts can be spread through the air, or other means, and alight on the surface of foodstuffs. Yeast colonies are generally moist or slimy in appearance and creamy white colored. Yeasts prefer an A_w of 0.90 - 0.94, but can grow below 0.90. These micro-organisms grow best in the intermediate acid range, pH from 4.0 to 4.5. Food that is highly contaminated with yeasts will frequently have a slightly fruity odour.

Viruses

Viruses are 10-450 nm in size; cannot reproduce without a living host; attack only susceptible host cell lines; infect plants, animals, and bacteria; and have the capacity to produce specific diseases in specific hosts. Transmission occurs in foods, water and air. Viruses that infect bacteria are called bacteriophages. Viruses are included in the order Virales.

Viruses are too small to be visualized with an ordinary compound microscope. Only after the electron microscope was developed, the direct observation of viruses was possible. Viruses consist of a DNA or RNA core surrounded by a protein coat. Because they lack all the apparatus for normal cellular metabolism, they must utilize the cellular machinery of the host cell in order to grow and divide. Once they invade a host cell, however, viruses can multiply very rapidly.

Parasitic Organisms

A number of parasitic worms can also be transmitted by food to cause diseases in humans. Cestodes are flatworms that inhabit the intestinal tract, heart, and lungs of animals. Beef, swine, dogs and other canine species, bears, and fish can all harbour tapeworms and flatworms, which can be transmitted to and can infect humans. Trematodes are non-segmented flatworms that possess a mouth and oral sucker and depend on a snail as an intermediate host before infecting humans by being ingested in drinking water or aquatic plants. Intestinal flukes, pyriform worms from fish, sheep and Chinese liver flukes, and oriental lung flukes are all examples of food-transmitted parasites.

Nematodes or true roundworms also can be transmitted from animals to humans. Eggs carried in excrement from roaches and dung beetles ingested by cattle, sheep and hogs contaminate humans. Trichinosis is an inflammation of the muscle tissue caused by ingesting the worm Trichinella spiralis. Pork is the most common vector. Capillary worms, whipworms, and pinworms are other examples of nematode parasites.

Protozoa are microscopic single-celled animals, which can be taken in with food or water to cause human illness. Entamoeba histolytica, Toxoplasma gondii, Balantidium coli, and Giardia lamblia are the most common food borne protozoan parasites.

Important Micro-Organisms in Food

Important Mold Genera

Molds are important in food because they can grow in conditions in which many bacteria cannot, such as low pH, low water activity (a_w), and high osmotic pressure. They are important spoilage micro-organisms. Many strains also produce mycotoxins and have been implicated in food borne intoxication. Many are used in food bioprocessing. Finally, many are used to produce food additives and enzymes. Some of the most common genera of molds found in food are listed here:

- Aspergillus: They are widely distributed and contain many species that are important in food. They have septate hyphae and produce a sexual spores (black color) or conidia. Many are xerophilic (able to grow in low Aw) and can grow in grains, causing spoilage. They are also involved in spoilage of foods such as jams, cured ham, nuts, and fruits and vegetables (rot). Some species/strains produce mycotoxin (e.g., Aspergillus flavus produces aflatoxin). Many species/strains are also used in food and food additive processing. Aspergillus oryzae is used to hydrolyze starch by alpha-amylase in the production of sake. Aspergillus niger is used to process citric acid from sucrose and to produce enzymes like-galactosidase.
- Alternaria: They are also septate and form dark-brown colored many celled conidia on the conidiophere. They cause rot in tomatoes and rancid flavor in dairy products. Species: Alternaria tenuis.

- Geotrichum: The hyphae are septate and form rectangular asexual arthrospores (oidia). They grow forming a yeast like, cottony, creamy colony. They establish easily in equipment and often grow on dairy products (also known as dairy mold). Species: Geotrichum candidum.
- Mucor: They are widely distributed. They have non-septate hyphae and produce sporangiophores. They produce cottony colonies. Some species are used in food fermentation and production of enzymes. They cause spoilage of vegetables. Species: Mucor rouxii.
- Penicillium: They are widely distributed and contain many species. They have septate hyphae and form conidiophore on a blue-green, brushlike conidia head. Some species are used in food production, such as Penicillium roquefortii and Penicillium camembertii in cheese. Many species cause fungal rot in fruits and vegetables.
- Rhizopus: The hyphae are aseptate and form sporangiophores in sporangium. They are involved in the spoilage of many fruits and vegetables. Rhizopus stolonifer is the common black bread mold.

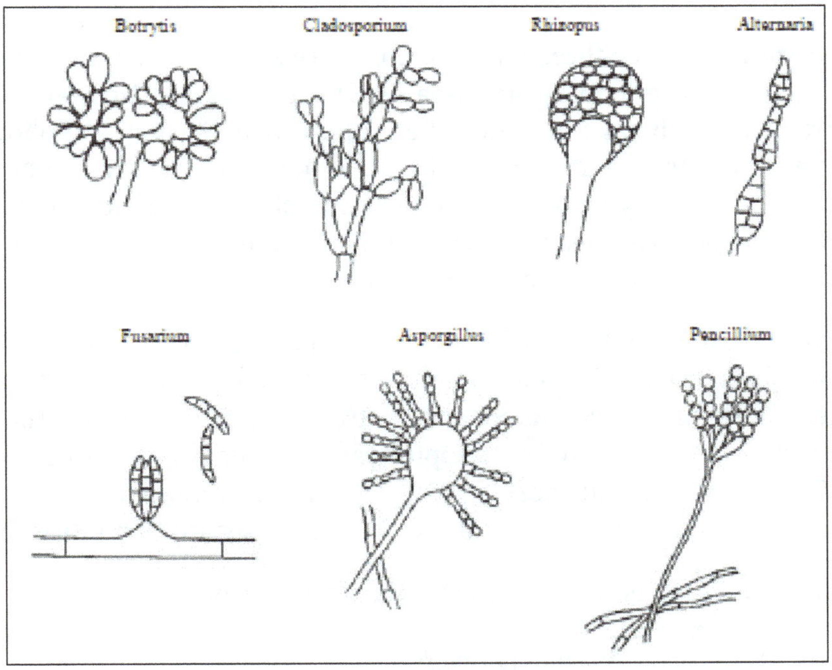

Common mold species found in foods.

Important Yeast Genera

Yeasts are important in food due to their ability to cause spoilage. Many are also used in food bioprocessing. Some are used to produce food additives. Several important genera are briefly described below:

- Saccharomyces: Cells are round, oval, or elongated. It is the most important genus and contains heterogeneous groups. Saccharomyces cerevisiae variants are used in baking for leavening of bread and in alcoholic fermentation. They are also involved in spoilage of food with the production of alcohol and CO_2.

- Pichia: They are oval to cylindrical cells and form pellicle in beer, wine, and brine to cause spoilage. Some are also used in oriental food fermentation. Species: Pichia membranaefaciens.
- Rhodotorula: They are pigment (red, pink or yellow) forming yeasts and can cause discoloration of foods, such as in meat, fish, and sauerkraut. Species Rhodotorula glutinis.
- Torulopsis: They have spherical to oval structure. They cause spoilage of milk due to the ability to ferment lactose (Torulopsis sphaerica). They also spoil fruit juice concentrates and acid foods.
- Candida: Many spoil foods with high acid, salt, and sugar and form pellicle on the surface of liquids. Some can cause rancidity in butter and dairy products (Candida lipolytica).
- Zygosaccharomyces: Involved in spoilage of foods, containing high sugar/ salt levels ex. honey, sirups, molasses, soy sauce. (Zygosaccharomyces nussbaumeri). These yeasts are termed osmophilic, because they can grow in high concentrations of solutes.

Important Viruses

Viruses are important in food for three reasons. Some are able to cause enteric disease and thus, if present in a food, can cause food borne diseases. Hepatitis A and Norwalk viruses have been implicated in food borne outbreaks. Several other enteric viruses, such as Poliovirus, Echovirus, and Coxsackievirus, have the potential of causing food borne diseases. In some countries where the level of sanitation is not very high, they can contaminate foods and cause disease. Some bacterial viruses (bacteriophages) are used in the identification of species/strains by a process called tranduction (e.g., in Escherichia, coli, Lactococcus lactis).

Finally, some bacteriophages can be very important due to their ability to cause fermentation failure. Many lactic acid bacteria, used as starter cultures in food fermentation, are sensitive to different bacteriophages. These phages can infect and destroy starter culture bacteria, causing product failure. Among the lactic acid bacteria, bacteriophages have been isolated for many species in genera Lactococcus, Streptococcus, Leuconostoc, and Lactobacills. Methods are being studied to genetically engineer lactic acid start cultures so that they become resistant to multiple bacteriophages.

Table: Human Intestinal Viruses with High Potential as Food Contaminants.

Types of Viruses	Example
Picornaviruses	Polioviruses Coxsackievirus A Coxsackievirus B Echovirus Enterovirus
Reoviruses	Reovirus Rotavirus
Parvoviruses	Human gastrointestional viruses
Papovaviruses	Human BK and JC viruses
Adenoviruses	Human adenoviruses

Important Bacterial Genera

Bacterial classification is rapidly changing. In the following table, only those species and genera currently approved and listed in Bergey's Manual have been used.

Table: Genera of Bacteria Important in Food.

Group	Family	Genera
Spiral and curved bacteria	Spirallaceae	Campylobacter
Gram-negative aerobic rods and cocci	Pseudomonadaceae	Pseudomonas, Altermonas, Gluconobacter, Xanthomonas
	Halobacteriaceae	Halobacterium, Halococcus
	Genera of uncertain affinity	Alcaligenes, Acetobacter, Brucella
Gram-negative facultative anaerobic rods	Enterobacteriaceae	Escherichia, Citrobacter Salmonella, Shigella, Klebsiella, Enterobacter, Serratia, Proteus, Yersinia, Erwinia
	Vibrionaceae	Vibrio, Aeromonas
	Genera of uncertain affinity	Flavobacterium, Chromobacterium
Gram-negative diplococcic and diplococcobacilli Gram-positive cocci	Neisseriaceae Micrococcaceae Streptococcaceae	Moraxella, Acinetobacter Micrococcus, Staphylococcus Streptococcus, Leuconostoc, Pediococcus, Lactococcus, Enterococcus
	Peptococcaceae	Sarcina
Endospore forming rods and cocci	Bacillaceae	Clostridium, Bacillus
Gram-positive asporogenous rod of regular shape	Lactobacillaceae	Lactobacillus
	Genera of uncertain affinity	Listeria
Non spore-forming rods of irregular shape	Coryneform bacteria	Arthrobacter, Brevibacterium, Propionibacterium
Rickettsia	Rickettsiaceae	Coxiella

Common Bacterial Groups in Foods

Among the Micro-organisms found in foods, bacteria constitute a major important group. This is not only because many different species can be present in foods, but is also due to their rapid growth rate, ability to utilize food nutrients, and their ability to grow under a wide range of temperatures, aerobiosis, pH, and water activity, as well as to survive under adverse situations, such as survival of spores at high temperature. For convenience, bacteria important in foods have been arbitrarily divided into several groups on the basis of similarities in certain characteristics. This grouping does not have any taxonomic significance. Some of these groups and their importance in foods are listed here:

- Lactic Acid Bacteria: Those bacteria that produce relatively large quantities of lactic acid from carbohydrates. Include species mainly from genera Lactococcus, Leuconostoc, Pediococcus, Lactobacillus and Streptococcus thermophilus.
- Acetic Acid Bacteria: Those bacteria that produce acetic acid, such as Acetobacter aceti.
- Propionic Acid Bacteria: Those bacteria that produce propionic acid and are used in dairy fermentation. Include species such as Propionibacterium freudenreichii.

- Butyric Acid Bacteria: Those bacteria that produce butyric acid in relatively large amounts. Some Clostridium spp., such as Clostridium butyricum.
- Proteolytic Bacteria: Those bacteria those are capable of hydrolyzing proteins due to production of extracellular proteinases. Species in genera Micrococcus, Staphylocccus, Bacillus, Clostridium, Pseudomonas, Alteromonas, Flavobacerium, and Alcaligenes; some in Enterobacteriaceae and Brevibacterium are also included in this group.
- Lipolytic Bacteria: Able to hydrolyze triglycerides due to production of extracellular lipases. Species in genera Micrococcus, Staphylococcus, Serration, Pseudomonas, Alteromonas, Alcaligenes and Flavobacterium are included in this group.
- Saccharolytic Bacteria: Able to hydrolyze complex carbohydrates. Include some species in genera Bacillus, Clostridium, Aeromonas, Pseudomonas, and Enterobacter.
- Thermophillic Bacteria: Able to grow at 50 oC and above. Include some species from genera Bacillus, Clostridium, Pediococcus, Streptococcus, and Lactobacillus.
- Psychrotrophic Bacteria: Able to grow at refrigerated temperature (<50). Include some species of Pseudomonas, Alteromonas, Alcaligenes, Flavobacterium, Serratia, Bacillus, Clostridium, Lactobacillus, Leuconostoc, Listeria, Yersinia and Aeromonas.
- Thermoduric Bacteria: Able to survive pasteurization temperature. Include some species of Micrococcus, Enterococcus, Lactobacillus, Pediococcus, Bacillus (spores) and Clostridium (spores).
- Halotolerant Bacteria: Able to survive high salt concentrations (>10%). Include some species of Bacillus, Micrococcus, Staphylococcus, Pediococcus, Vibrio Streptococcus, Clostridium and Corynebacterium.
- Aciduric Bacteria: Able to survive at low pH (below 4.0). Include some species of Lactobacillus, Pediococcus, Lactococcus, Enterococcus and Streptococcus.
- Osmophilic Bacteria: Can grow at a relatively higher osmotic pressure (environment) than other bacteria. Some species from genera Staphylococcus, Leuconostoc, and Lactobacillus are included in this group. They are much less osmophilic than yeasts and molds.
- Gas-producing Bacteria: Produce gas (CO_2, H_2, H_2S) during metabolism of nutrients. Include spices from genera Leuconostoc, Lactobacillus, Brevibacterium and Escherichia.
- Slime Producers: Produce slime due to synthesis of polysaccharides. Include some species or strains of Xanthomonas, Leuconostoc, Alcaligenes, Enterobacter, Lactococcus, and Lactobacillus.
- Spore formers: Ability to produce spore. Include Bacillus, Clostridium and Desulfotomaculum spp. They are again divided into aerobic, anaerobic, flat sour thermophilic and sulfide-producing spore formers.
- Aerobes: Require oxygen for growth and multiplication. Species of Pseudomonas, Bacillus, and Flavobacterium are included in this group.
- Anaerobes: Cannot grow in the presence of oxygen. Include species of Clostridium.

- Facultative Anaerobes: Able to grow both in the presence and absence of oxygen. Lactobacillus, Pediococcus, Leuconostoc, enteric pathogens, some species of Bacillus, Serratia, and coliforms are included in this group.
- Coliforms: Include mainly species from Escherichia, Enterobacter, Citrobacter, and Klebsiella, and used as index of sanitation.
- Fecal Coliforms: Include mainly Escherichia coli. Also used as index of sanitation.
- Enteric Pathogens: Include pathogenic Salmonella, Shigella, Campylobacter, Yersinia, Escherichia, Vibrio, Listeria, Hepatitis A, and others that can cause gastrointestinal infection.

Normal Micro Flora of Some Common Foods

Under normal conditions a food generally harbours only a few types of Microorganisms. They constitute those that are naturally present in raw foods (which provide the ecological niche) and those that get in from outside sources to which the foods are exposed from the time of production until consumption. The predominant type(s) will be the ones for which the optimum growth condition is present.

Meat

The carcass of a healthy animal slaughtered for meat and held in a refrigerated room is likely to have only nominal surface contamination while the inner tissues are sterile. Fresh meat cut from the chilled carcass has its surface contaminated with micro-organisms characteristic of the environment and the implements (saws or knives) used to cut the meat. Each new surface of meat, resulting from a new cut, adds more micro-organisms to the exposed tissue.

Among the most common species of bacteria occurring on fresh meats are Pseudomonads, Staphylococci, Micrococci, Enterococci and Coliforms. The low temperature at which fresh meats are held favors the growth of psychrophilic Micro-organisms.

Poultry

Freshly dressed eviscerated poultry have a bacterial flora on their surface (skin) that originates from the bacteria normally present on the live birds and from the manipulations during killing, defeathering, and evisceration. Under good sanitary conditions the bacterial count has been reported to be from 100 to 1000 bacteria per square centimeter of skin surface, whereas under less sanitary conditions the count may increase 100-fold or more. Pseudomonads constitute the major contaminants on the skin of freshly dressed poultry.

Eggs

The interior of a freshly laid egg is usually free of micro-organisms; its subsequent microbial content is determined by the sanitary conditions under which it is held, as well as the conditions of storage, i.e. temperature and humidity. Micro-organisms particularly bacteria and molds, may enter the egg through cracks in the shells or penetrate the shells when the "bloom" (thin protein coat) covering the shell deteriorates. The type of micro-organisms involved reflect those present in the environment.

Fruits and Vegetables

They are normally susceptible to infection by bacteria, fungi, and viruses. Microbial invasion of plant tissue can occur during various stages of fruit and vegetable development, and the likelihood of spoilage increases. A second factor contributing to the microbial contamination of fruits and vegetables pertains to their post-harvest handling. Mechanical handling is likely to produce breaks in the tissue which facilitates invasion by micro-organisms. The pH of fruits is relatively acid i.e. ranging from 2.3 (for lemons) to 5.2 (for bananas). This restricts bacterial growth but does not retard fungal growth. The pH range for vegetables is slightly higher pH 5.0 to 7.0 and hence they are more susceptible than fruits to bacterial attack.

Shellfish and Finfish

The microbial flora of freshly caught oysters, clams, fish, and other aquatic specimens is very largely a reflection of the microbial quality of the water from where they are harvested. Of particular significance is whether the water is sewage-polluted, in which case the aquatic food is potentially capable of transmitting various pathogenic Micro-organisms. Shellfish that grow in contaminated water can concentrate viruses and may be the source of Hepatitis infection. For example, raw oysters and clams from polluted waters have caused numerous epidemics in various parts of the world.

Milk

Milk is an excellent growth medium for all of the common spoilage organisms, including molds and yeasts. Fresh, non-pasteurized milk generally contains varying numbers of Micro-organisms, depending on the care employed in milking, cleaning, and handling of milk utensils. Raw milk held at refrigerator temperatures for several days invariably shows the presence of several or all bacteria of the following genera: Enterococcus, Lactococcus, Streptococcus, Leuconostoc, Lactobacillus, Microbacterium, Propionibacterium, Micrococcus, Coliforms, Proteus, Pseudomonas, Bacillus, and others. Those unable to grow at the usual low temperature of storage tend to be present in very low numbers. The pasteurization process eliminates all but thermoduric strains, primarily, Streptococci and Lactobacilli, and spore formers of the genus Bacillus (and clostridia if present in raw milk). The spoilage of pasteurized milk is caused by the growth of heat-resistant Streptococci utilizing lactose to produce lactic acid, which depresses the pH to a point (about pH 4.5) where curdling takes place.

Microbial Growth

Microbial growth dynamics is a subject of numerous fundamental and applied research studies in modern microbiology and biotechnology. Usually, biotechnologists want to know the time prog-ress of, say, product formation associated with cell growth, nutrients uptake, respiration, and oth-er metabolic processes. We can predict growth dynamics before setting up a real experiment or technological line by using mathematical simulation models of different degrees of complexity. Such a preliminary simulation is very useful for planning and optimizing real-life experiments. Moreover, the verified dynamic mathematical models provide efficient tools to optimize the yield of target product, minimize undesirable generation of waste products, etc.

Another important implication of growth dynamics is that recorded growth curves carry important hidden information about microbial cells, growth regulation, and interactions. Based on growth dynamic pattern, we can distinguish the effects of products or substrate inhibition, identify growth-limiting substrates at various stages, pinpoint the importance of positive or negative interactions in a mixed or genet-ically inhomogeneous culture, etc.

The scientific discipline that uses special quantitative tools to study the development of any processes in time – physical chemical, or biological – is called kinetics. Kinetic studies in microbiology cover all dynamic manifestations of microbial life: growth itself, survival and death, product formation, adaptations, mutations, cell cycles, environmental effects, and biological interactions. Kinetics provide a theoretical framework for optimal design in biotechnologies, based on fermentation and enzyme catalysis, as well as on the employment of outdoor activity of natural microbial populations (wastewater treatment, soil bioremediation, etc).

Contrary to simple rate measurements, kinetic studies require the perception of the underlying basic mechanisms of studied processes. We will define mechanistic studies as those that interpret some complex process as an interplay of several simpler reactions; for example, cell growth can be explained through the activity of enzymes and microbial community dynamics can be interpreted through the behaviour of individual cells and populations. Ideally, mechanistic studies infer the coupling of experimental measurements with analysis of simulating mathematical models. The models formalize postulated mechanisms, so that the comparison of observations and the model's predictions allows one to discard an incorrect hypothesis.

Quantitative studies in microbiology often involve the assessment of 'growth stoichiometry'. Stoichiometry is the quantitative relationship between reactants and products in a chemical reaction. In microbiology, stoichiometry stands for a quantitative relationship between substrates and products of microbial processes, including biomass formation (the consequence of complying with mass and energy conservation laws). In practical terms, kinetic and stoichiometry are tightly linked to each other, but stoichiometry mainly addresses problems of a static nature (how much? in what proportion?), whereas kinetics considers the dynamics questions (at what rate? by which mechanism?).

Kinetic Models of Microbial Growth

The description of microbial growth depends critically on a 'language' used to convey growth-related information. Such language is called 'kinetic model'. It can be very simple and represent the growth of cells as a uniform exponential function of time (all other non-exponential phases of growth being ignored). On the other extreme, a model can describe cell growth as a cumulative result of thousands of individual metabolic reactions generated from a genome sequence of a particular organism and based on balancing of metabolic fluxes. Some complex models are designed to describe spatially organized microbial colonies, biofilms, or mycelial network colonizing soil or other natural habitats. In this section, we will overview the most popular biotechnology growth models with different degrees of complexity, starting from the so-called non-structured models, which treat microbial growth as an autocatalytic reaction:

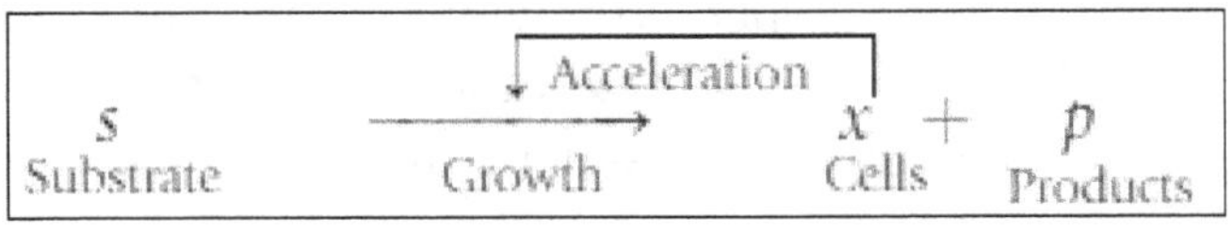

In the non-structured models, we ignore any changes in the quality of grown cells, which arise mainly through differential gene expression (spectrum of expressed enzymes, chemical composition of cells, size of the metabolic pool, shape and size of cells, etc.). More realistic structured models that get inside the cells will be considered later together with non-compromised systems biology approaches stemming from the omics data.

State Variables and Growth Parameters

The basic state variables of simple growth models are listed in Table. For unicellular microorganisms (the majority of bacteria and yeasts), we use two measures of cellular abundance, the concentration of cell mass (x) and cell number (N). They are related to each other by a simple formula:

$$x = mN$$

Where, conversion factor m is the average dry mass of a single cell. In some studies, it is safe to assume m to be a constant and use both variables x and N as equivalent measures of growth. However, the average size and mass of single cells vary, depending on the nature of studied organisms and environmental conditions; therefore, it is advisable to make a selection. The biomass, x, has an obvious advantage in studies related to mass and energy balance of microbial growth, while the cell number, N, is preferred in population and genetic studies when, for example, mutation or plasmid transfer is an essential factor controlling the efficiency of the studied process.

Table: Major state variables of growth models.

Variable	Notation	Dimension (examples)
Concentration of cell biomass	x	g-CDW l^{-1} of cultural liquid
Cell number	N	10^9 cell ml^{-1} of cultural liquid
Single-cell mass	m_1	g-CDW per cell
Mycelium length	L	ml^{-1} of cultural liquid
Tips number	n	10^6 tips ml^{-1} of cultural liquid
Concentration of limiting substrate	s	g l^{-1} of cultural liquid
Concentration of product	p	g l^{-1} of cultural liquid or g g^{-1} of CDW

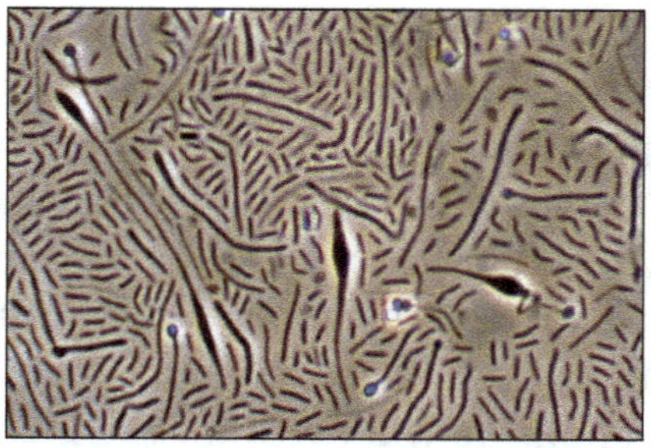

Example of spontaneous variation in size and shape of bacterial cells. Clostridium thermocellum ATCC 27405 grown anerobically in batch culture on cellobiose.

The choice of method to measure x or N depends on many factors. Today, the preference is given to automated fast and precise techniques. The most advanced analytical methodology (Fourier transform infrared spectroscopy (FTIR), mass spectrometry, and nuclear magnetic resonance (NMR)) is now available for automatic recording of numerous microbial products and nutrient substrates, especially if they are represented by gases or volatile compound, such as methane, CO_2, O_2, H_2, fatty acids, alcohols, and other fermentation products.

Table: Methods used for experimental determination of microbial cells.

Method	Measuring principle	DL	Advantages	Disadvantages
Dry Dry mass	Mass of separated and dried solids	50	Provides direct unconditional estimate	Interference from dead cells and non-cell solids
Wet mass	Mass of separated material	50	Simplicity, quickness	The variation of wet bio-mass bulk density
Wet bio-vol-ume	Linear dimension of pelleted cells or colony	20–100	Simplicity, quickness	The variation of wet bio-mass bulk density
Particulate or-ganic carbon	CO_2 after cell separation and combustion	0.1–1.0	High precision and sensitivity, provides direct estimate of mass	Interference from dead cells and non-cell solids
Biuret pro-teins	Colorimetric reaction of peptide bonds	1.0	High uniformity	Variation of protein-to-bio-mass ratio, possible extra-cellular accumulation
Folin-Ciocal-teu protein	Colorimetric reaction of tyrosine and tryptophan	0.1	High sensitivity	Variation of protein-to-bio-mass ratio and amino acid composition of cell protein, possible extracellular accu-mulation
DNA	Colorimetric estimation of deoxyribose	1.0	High specificity, constancy of the DNA cellular content	Possible extracellular accu-mulation
ATP	Bioluminescence assay	10^{-5}	High sensitivity and specificity	Variation of the intracellular ATP pool
Fatty acids	Gas or liquid chroma-tography, colorimetric reaction	1.0	High specificity, allows to identi-fy composition of mixed culture	Variation of the intracellular content, possible extracellu-lar accumulation
Metabolic potential	Rate of added substrate uptake or product for-mation	10^{-3} –10	High specificity, quickness	Variation in conversion factor from measured rate to biomass
Turbidity	Light scattering	0.1	Simplicity, quickness, automa-tion	Interference from non-cell solids, cell aggregation, wall growth
Electrical measurements	Conductivity	1	Simplicity, quickness, automa-tion	Variation of conversion factor, interference from electrolytes in the medium
Manual mi-croscopy	Cell count, measuring of hypha length	10^{-5}	Low cost, sensitivity, allows vi-sual assay of bio-morphological structure	Time consuming, low pre-cision
Comput-er-aided microscopy (image anal-ysis)	Computer-aided count	10^{-5}	Combined benefit of manual and automated quantification, speed, generation of size distribution	High cost
Coulter count-er	Automated count and sizing	10^{-5}	Quickness, automation, genera-tion of size distribution	Interference from non-cell solids
Plating	Growing of the colonies on Petri dish	10^{-5}	Low cost, high sensitivity	Time consuming, low pre-cision

The primary state variables (concentration of cell mass, x, substrate, s, and product, p, with subscript zero indicating respective initial values) that are measured directly in cell culture are usually recalculated into the secondary growth characteristics sometimes called parameters:

- Gross growth rate: $\frac{dx}{dt}$
- Specific growth rate: $\mu = \frac{1}{x}\frac{dx}{dt}$
- Degree of multiplication: $\frac{x}{x_o}$
- Biomass doubling time: $t_d = \ln\frac{2}{\mu}$
- Growth yield: $Y = \frac{-dx}{ds} \sim \frac{(x - x_0)}{(s_0 - s)}$
- Product yield: $Y_{p/s} = \frac{-dp}{ds}$ or $Y^{p}/_{x} = \frac{dp}{dx}$
- Specific rate of substrate consumption: $q_s = -\frac{\left(\frac{ds}{dt}\right)}{x} = \frac{\mu}{Y}$
- Specific rate of product formation: $q_p = \frac{\left(\frac{dp}{dt}\right)}{x} = \mu Y_{p/x} = \mu\left(\frac{Y_{p/x}}{Y}\right)$

Most of the listed secondary growth characteristics are specific rates, expressing microbial activities per unit of cell mass. Specific rates have preference because they are not sensitive to the amount of cell biomass, an extremely variable entity. The specific growth rate, μ, measured from biomass dynamic, x, can differ from that estimated from the increase in cell number, N (denoted as μ_N). It follows from equationn $x = mN$ that,

$$\mu = \frac{1}{x}\frac{dx}{dt} = \frac{1}{x}\frac{d(Nm)}{dt} = \frac{1}{Nm}\left(N\frac{dm}{dt} + m\frac{dN}{dt}\right) = \frac{1}{m}\frac{dm}{dt} + \frac{1}{N}\frac{dN}{dt} = \mu_{cell} + \mu_N$$

If mass of single cell m is constant, then $\mu_{cell} = 0$ and $\mu = \mu_N$. Otherwise, we have to take into account m-variation.

Exponential Growth

Validity of exponential growth law. One of the earliest postulates in microbial kinetics is that, under optimal non-restrictive conditions (nutrient media containing all essential components at non-limiting concentrations, absence of inhibitors, adequate physicochemical parameters, and perfect mixing), the increase in biomass (dx) during an infinitely small time interval (dt) is proportional to dt and the instant biomass concentration (x), that is,

$$dx = \mu x\, dt \text{ or} \frac{dx}{dt} = \mu x$$

where μ is the proportionality coefficient. If μ is constant, integration of equation above gives:

$$\ln x = \ln x_0 + \mu t \Rightarrow x = x_0 \exp(\mu t)$$

where, x_0 is the biomass at zero time or inoculum size.

According to early views, true exponential growth occurs only in the case of symmetrically dividing bacteria with equal probability of subsequent division for the mother and daughter cells. Organisms with asymmetrical multiplication (budding yeasts) were thought to obey the exponential law only approximately, whereas the growth of filamentous organisms (fungi and streptomycetes) was deviated considerably. However, direct measurements performed from 1930 to 1960 revealed that exponential growth does occur in all prokaryotes and eukaryotes independent of their bio-morphological features, including protozoa, fungi, and homogeneously cultivated plant and animal cells. The deviation is observed only as a result of growth-associated changes in the environment. Wang and Koch made very precise measurements of μ dynamics by growing Escherichia coli directly in an aerated cuvette of the computer-linked double-beam spectrophotometer. They found temporary slowdowns in the complex beef extract media attributed to the diauxie phenomena (e.g., depletion of some preferred oligopeptides) and a gradual increase of the μ value during sequential subculture in succinate minimal medium. The derivation of the exponential growth equation was done for binary dividing bacteria, based on the geometric progression 2, 2^2, 2^3:

$$N = N_0 2^n$$

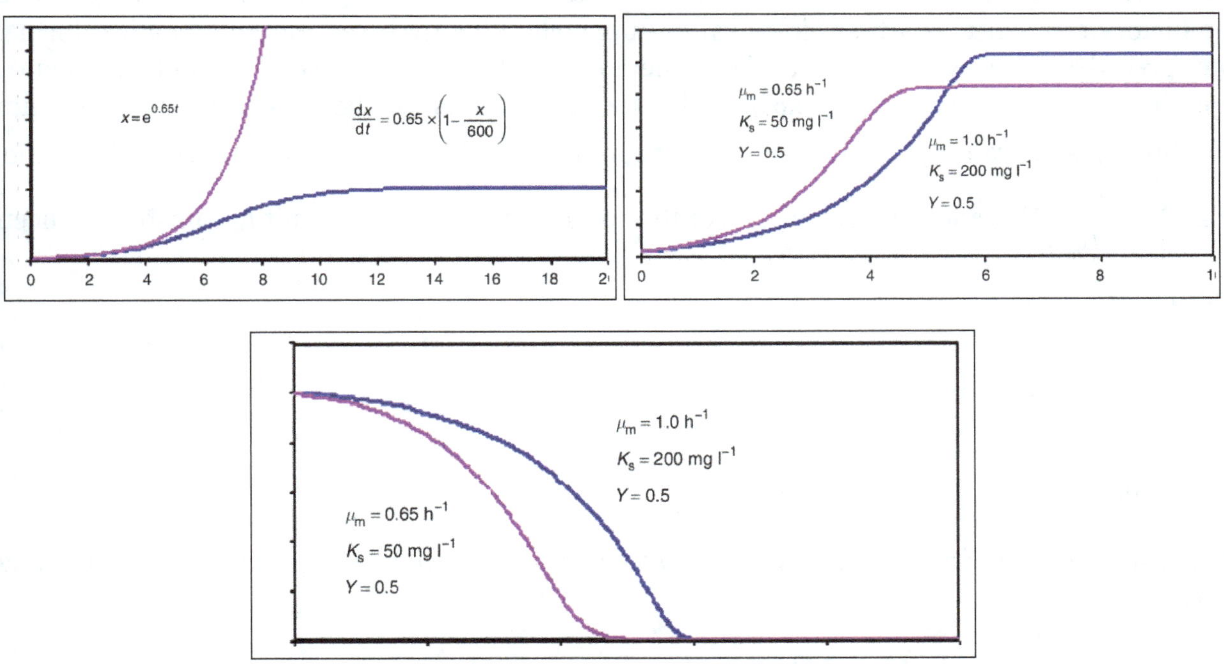

Simple growth curves: Exponential and logistic equations (top panel), Monod model for cell dynamics (middle panel), and residual limiting substrate (bottom panel).

Where, n is the number of divisions, $n = t/t_d = \mu t/\ln 2$. It is easy to show that 2^n is equivalent to expression $e^{\mu t}$ for any type of division, for example, when the mother cell divides into 2, 3, or any other number of offspring, the only condition to be met being that the propagation pattern must remain constant through time.

To accommodate the explosive exponential growth with real constrained cell dynamics, several empirical adjustments were used. For instance, batch growth was viewed as a series of exponential phases, each of which has a unique value of empirical function F(t):

$$N = N_0 \exp\left[\mu F(t) t\right]$$

Where F(t) switches from 0 to 1.0, again 0, and then to −1 in the lag, log, stationary and death phases, respectively. Another empirical way to correct the exponential term was the use of a logistic equation:

$$\frac{dx}{dt} = rx - r\frac{x^2}{x_m} = rx\left(1 - \frac{x}{x_m}\right)$$

On the right-hand side, the exponential term rx (r is the difference between true growth and death rates) is compensated for by a term describing the negative interactions between organisms, which are proportional to x^2 and inversely related to x_m, the upper density limits (the carrying capacity, K, in ecological literature). This equation predicts almost exponential growth at low density x with subsequent slowing down while asymptotically approaching x_m.

Monod Model: Explanation of Growth Limitation by Availability of Substrate

This model is still very popular because of its elegant simplicity. It played an important role in the history of science as a first biological example when theory based on mathematical formalisms preceded novel experimental designs (chemostat culture). Contrary to the simple exponential equation $\ln x = \ln x_0 + \mu t \Rightarrow x = x_0 \exp(\mu t)$, this model accounts for the mass-conservation condition linking substrate uptake to biomass formation through a constant yield factor, Y equation $Y = \frac{-dx}{ds} \sim \frac{(x - x_0)}{(s_0 - s)}$, and the dependence of the specific growth rate on limiting substrate concentration:

$$\frac{dx}{dt} = \mu(s)x = \mu_m \frac{s}{K_s + s} x$$

$$\frac{ds}{dt} = -\frac{1}{Y}\frac{dx}{dt}$$

By the substitution of s by x via Y, we can reduce equation above to a single ODU having the analytical solution:

$$\mu_m t = (1 + P)\ln\left(\frac{x}{x_0}\right) - P\ln\left(\frac{Q - x}{x_0}\right) + P\ln(Q - 1)$$

where, $P = YK_s/x_m$, $Q = x_m/x_0$, $x_m = Ys_0 + x_0$, while s_0 and x_0 are, respectively, s and x at t = 0.

Equation $\frac{dx}{dt} = \mu(s)x = \mu_m \frac{s}{K_s + s} x$, $\frac{ds}{dt} = -\frac{1}{Y}\frac{dx}{dt}$ contains three parameters, Y, μ_m, and K_s, that can be thought of as a personal ID for any particular organism. If these parameters are known and the initial conditions, so and xo, set up by the experimenter are also known, and then $\frac{dx}{dt} = \mu(s)x = \mu_m \frac{s}{K_s + s}x, \ \frac{ds}{dt} = -\frac{1}{Y}\frac{dx}{dt}$ allow the calculation of cell mass and residual substrates prior to the actual experiment. Jacque Monod demonstrated excellent agreement of his data on E. coli with the described model (although not many other researchers were able to confirm it), but even more important is that this model led to the development of the chemostat theory, the set of ordinary differential equations (ODEs) describing bacterial propagation in a continuous-flow bioreactor before respective hardware was constructed. For this purpose, equations $\frac{dx}{dt} = \mu(s)x = \mu_m \frac{s}{K_s + s}x, \ \frac{ds}{dt} = -\frac{1}{Y}\frac{dx}{dt}$ were modified as follows:

$$\frac{dx}{dt} = (\mu - D)x, \quad \mu = \mu_m \frac{s}{K_s + s}x$$

$$\frac{ds}{dt} = D(s_r - s) - \frac{1}{Y}\mu x, \quad D = \frac{F}{V}$$

Where s_r is the concentration of limiting substrate in the fed-fresh medium delivered with a pump from a reservoir. F is the pumping rate ($cm^3\ h^{-1}$), V is the culture volume (cm^3), and D is the dilution rate defined as the ratio D = F/V (h^{-1}).

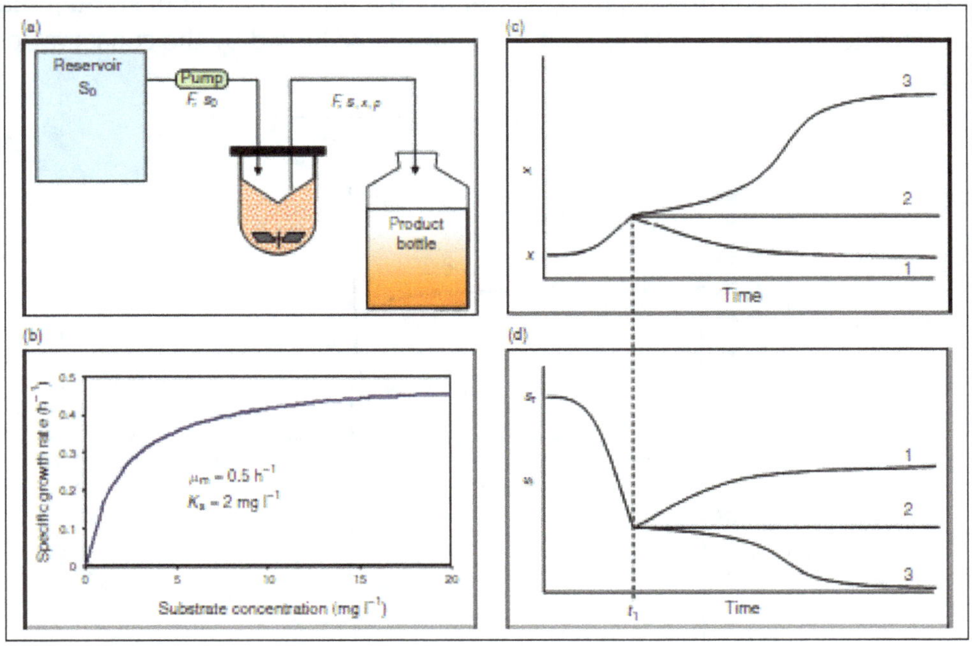

Chemostat operation

Figure shows (a) diagrammatic representation; (b) specific growth rate (μ) as a function of limiting substrate concentration, s, the only operating negative feedback mechanism stabilizing continuous culture; (c) the three possible outcomes affecting cell mass concentration in bioreactor (x): (1) rate of cell washout exceeds maximum growth rate; (2) rate of washout = growth rate, and (3) initial washout rate is less than growth rate; and (d) the same three outcomes affecting concentration of limiting substrate (s).

The mathematical analysis of equation above leads to several nontrivial conclusions regarding the biological nature of continuous growth:

1. The described open system attains stable steady state, when dx/dt = 0, ds/dt = 0, and variables x and s are not changed with time (denoting steady-state concentrations $\tilde{x}$ and $\tilde{s}$, respectively). From the first equation, it follows that μ = D, that is, the specific rate of microbial growth is equal to the dilution rate, which is under the full control of the experimenter. From the second equation, we find,

$$\tilde{x}=Y(s_0-\tilde{s})$$

As compared with the Y expression for batch culture (equation $Y=\frac{-dx}{ds}\sim\frac{(x-x_0)}{(s_0-s)}$), this formula no longer contains the term, x_0, the inoculum size. Thus, a steady-state culture is not dependent on history, and its properties are determined solely by current cultivation parameters.

2. The dilution rates, D, permitting stable microbial growth, is confined between the limits 0 and washout point or critical dilution rate, D_c:

$$D_c=\frac{\mu_m s_r}{(K_s+s_r)}$$

3. The specific growth rate μ does not depend on either so or x and is governed solely by the substrate concentration in the culture. On the other hand, the indirect effect of so on μ may be evaluated from the dependence of $\tilde{x}$ on D, as soon as $\tilde{x}$ and $\tilde{s}$, re linked by the conservation condition in equation $\tilde{x}=Y(s_0-\tilde{s})$.

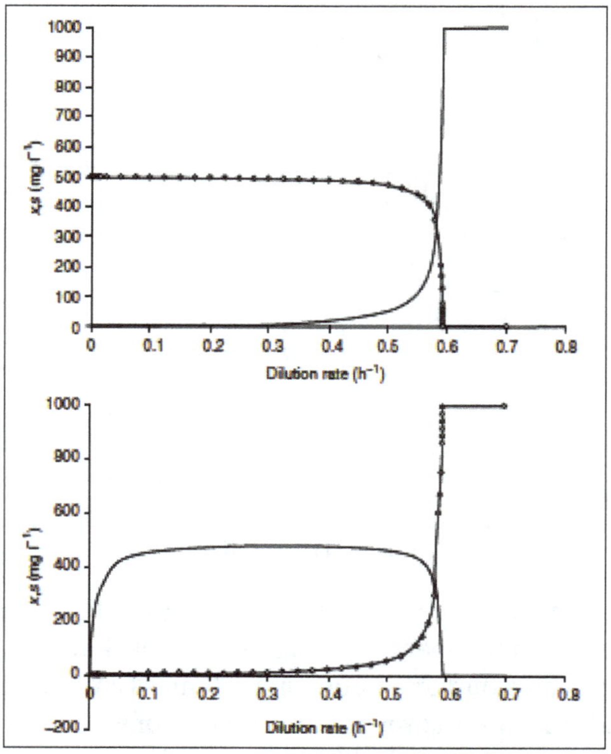

Steady-state values of biomass (◦) and growth-limiting substrate concentrations (—) in a chemostat. Top: No maintenance requirements (m=0). Bottom: m = 20mg h^{-1} g^{-1} DW.

The most important biological implication of the chemostat theory was the discovery of the negative feedback between microbial growth rate and concentration of substrate in the bioreactor to establish steady state. The second surprising fact was that microorganisms can grow endlessly with any rate between 0 and μ_m. Such substrate-limited growth is nevertheless exponential, because two subsequent acts of cell division are separated by a constant time interval. Earlier, substrate limitation had been observed only transiently at the end of the exponential phase of batch growth. The most important biotechnological implication of the chemostat model was the concept of controlled cell biosynthesis, which implies the purposeful manipulation of microbial culture to optimize the productivity by selecting the most appropriate flow rate and by changing the composition of medium in continuous flow bioreactors.

Biological meaning and experimental determination of growth parameters K_s and μ_m. The parameter μ_m, maximal specific growth rate, has a very important biological meaning: it is the upper limit of μ variation in a specified nutrient medium. It could not be attained in reality because of its asymptotic nature: $\mu \rightarrow \mu_m$ as $s \rightarrow \infty$. However, in practice, μ_m is achieved if s >> K_s. It should be remembered that μ_m is not an absolute maximum of growth rate, as it depends on the nature of the limiting substrate. For example, E. coli has a μ_m above 2.5 h^{-1} on complex beef-extract medium and below 1.0 h^{-1} on minimal medium with succinate. Saturation constant, K_s, is numerically equal to substrate concentration which supports $0.5\mu_m$. We can say also that K_s, is a measure of cell affinity to substrate: the lower the K_s, the better is the organism adapted to absorb substrates from diluted solution. Any other definitions are speculative; for example, K_s is interpreted as the dissociation constant of ESC of the cellular enzyme involved in the first step of substrate conversion.

There are several ways to determine the numeric values of K_s and μ_m:

- Batch culture: By using an appropriate plate reader, one has to follow the biomass dynamics, x(t), from the start of the exponential phase to the point of complete substrate consumption (we should know so) when cell density reaches the maximum (x = xm). The residual substrate concentration, s(t), is calculated from mass balance (equation $Y = \frac{-dx}{ds} \sim \frac{(x - x_0)}{(s_0 - s)}$) and corresponding μ(t) values are found by the numeric differentiation of x(t) curves by applying the standard SLOPE function (μ = d(ln x)/dt) in MS Excel. Finally, the parameters K_s and μ_m are found by the nonlinear regression (Solver in Excel) of μ versus s. An alternative way is to fit immediately the growth curve to (equation $\mu_m t = (1+P)\ln\left(\frac{x}{x_0}\right) - P\ln\left(\frac{Q-x}{x_0}\right) + P\ln(Q-1)$), the integrated form of the Monod equation. Sometimes (e.g., when cells are grown in opaque media), it is more convenient to estimate Ks by recording residual substrate, s(t), or product formation, p(t), for example, CO_2 evolution, or O_2 uptake.
- Steady-state chemostat culture: The chemostat is run at several dilution rates, D, the corresponding $\tilde{s}$ values are measured, and, therefore, the dependence of μ = D on $\tilde{s}$ can be obtained, in principle, as accurately and carefully as desired. Such an approach, however, may and frequently does encounter serious technical problems because of the high affinity to the limiting substrate of some microorganisms. It is important to select sensitive

analytical techniques to measure the residual substrate and to develop instant sampling procedures (good results are obtained with the tangential flow) under intensive stirring to minimize the nutrient–substrate mixing time. The use of radioisotopes provides appropriate sensitivity; some other valuable practical advices are given by Button.

- Non-steady-state chemostat culture: The measurements are made during short-term spike experiments by the addition of different amounts of limiting substrate to a steady-state chemostat culture, directly to the main vessel or in a circulation loop, followed by recording substrate-stimulated dissolved oxygen (DO) uptake. Contrary to the steady-state method, this technique is much faster, more sensitive, and is able to show progressive changes in uptake kinetics as dependent on dilution rate in chemostat.

Washout method to assess maximal specific growth rate: The steady-state chemostat culture is perturbed by setting the dilution rate 20–100% higher than the critical value D_c (equation $D_c = \frac{\mu_m s_r}{(K_s + s_r)}$. The washout dynamics is followed and μm is determined from the approximate relationship: $x \approx x(0)\exp[(\mu_m - D)t]$, where x(0) is the biomass just before perturbation. A more rigorous estimate of μm could be provided if such washout experiments are made at several (at least two) input ubstrate concentrations, s_r, to account for the fact that s_r is still not infinity.

Chapter 2

Microbial Food Spoilage

The process which makes the food undesirable and unacceptable for consumption is known as food spoilage. Some of the important topics studied under food spoilage are meat and poultry spoilage, seafood spoilage, fruit and vegetable spoilage, and milk and dairy products' spoilage. This chapter has been carefully written to provide an easy understanding of these varied facets of food spoilage.

Food Spoilage

Food spoilage is a metabolic process that causes foods to be undesirable or unacceptable for human consumption due to changes in sensory characteristics. Spoiled foods may be safe to eat, i.e. they may not cause illness because there are no pathogens or a toxin present, but changes in texture, smell, taste, or appearance cause them to be rejected. Some ecologists have suggested these noxious smells are produced by microbes to repulse large animals, thereby keeping the food resource for themselves. Food loss, from farm to fork, causes considerable environmental and economic effects. The USDA Economic Research Service estimated that more than ninety-six billion pounds of food in the U.S. were lost by retailers, foodservice and consumers. Fresh produce and fluid milk each accounted for nearly 20% of this loss while lower percentages were accounted for by grain products (15.2%), caloric sweeteners (12.4%), processed fruits and vegetables (8.6%), meat, poultry and fish (8.5%), and fat and oils (7.1%). Some of this food would have been considered still edible but was discarded because it was perishable, past its sell-by date, or in excess of needs. There are also environmental and resource costs associated with food spoilage and loss. If 20% of a crop is lost, then 20% of the fertilizer and irrigation water used to grow that crop was also lost. Shelf life of a food is the time during which it remains stable and retains its desired qualities.

The wide array of available dairy foods challenges the microbiologist, engineer and technologist to find the best ways to prevent the entry of micro-organisms, destroy those that do get in along with their enzymes, and prevent the growth and activities of those that escape processing treatments. Troublesome spoilage micro-organisms include aerobic psychrotrophic Gram-negative bacteria, yeasts, molds, hetero-fermentative lactobacilli, and spore-forming bacteria. Psychrotrophic bacteria can produce large amounts of extracellular hydrolytic enzymes, and the extent of recontamination of pasteurized fluid milk products with these bacteria is a major determinant of their shelf life. Fungal spoilage of dairy foods is manifested by the presence of a wide variety of metabolic by-products, causing off-odors and flavors, in addition to visible changes in color or texture.

Coliforms, yeasts, hetero-fermentative lactic acid bacteria, and spore-forming bacteria can all cause gassing defects in cheeses. The rate of spoilage of many dairy foods is slowed by the application of one or more of the following treatments: reducing the pH by fermenting the lactose to lactic acid; adding acids or other approved preservatives; introducing desirable microflora that restricts the growth of undesirable microorganisms; adding sugar or salt to reduce the water activity (a_w); removing water; packaging to limit available oxygen; and freezing. The type of spoilage microorganisms differs widely among dairy foods because of the selective effects of practices followed in production, formulation, processing, packaging, storage, distribution, and handling.

Scenario of Food Spoilage Worldwide

The issue of food losses is of high importance in the efforts to combat hunger, raise income and improve food security in the world's poorest countries. Food losses have an impact on food security for poor people, on food quality and safety, on economic development and on the environment. The exact causes of food losses vary throughout the world and are very much dependent on the specific conditions and local situation in a given country. In broad terms, food losses will be influenced by crop production choices and patterns, internal infrastructure and capacity, marketing chains and channels for distribution, and consumer purchasing and food use practices. Irrespective of the level of economic development and maturity of systems in a country, food losses should be kept to a minimum.

Food losses represent a waste of resources used in production such as land, water, energy and inputs. Producing food that will not be consumed leads to unnecessary CO_2 emissions in addition to loss of economic value of the food produced. Economically avoidable food losses have a direct and negative impact on the income of both farmers and consumers. Given that many smallholders live on the margins of food insecurity, a reduction in food losses could have an immediate and significant impact on their livelihoods. For poor consumers (food insecure or at-risk households), the priority is clearly to have access to food products that are nutritious, safe and affordable.

It is important to note that food insecurity is often more a question of access than a supply problem. Improving the efficiency of the food supply chain could help to bring down the cost of food to the consumer and thus increase access. Given the magnitude of food losses, making profitable investments in reducing losses could be one way of reducing the cost of food. But that would, of course, require that financial gains from reduced losses are not outweighed by their costs. How much food is lost and wasted in the world today and how can we prevent food losses? Those are questions impossible to give precise answers to, and there is not much on-going research in the area. This is quite surprising as forecasts suggest that food production must increase significantly to meet future global demand.

Worldwide postharvest fruit and vegetables losses are as high as 30 to 40% and even much higher in some developing countries. Reducing postharvest losses is very important; ensuring that sufficient food, both in quantity and in quality is available to every inhabitant in our planet. The prospects are also that the world population will grow from 5.7 billion inhabitants to 8.3 billion in 2025. World production of vegetables amounted to 486 million ton, while that of fruits reached 392 million ton. Reduction of post-harvest losses reduces cost of production, trade and distribution, lowers the price for the consumer and increases the farmer's income.

Fruits and vegetables are very important food commodities not only in India but all over the world. India, which is the second most populated country of the world, is still struggling to achieve self-sufficiency to feed about 800 million people. For this purpose, fruits and vegetables have got their specific importance to provide a balance and healthy diet to the people. India is the second largest producer of vegetables and fourth largest producer of fruits in the world. Though India is producing adequate quantities of fruits and vegetables, yet on account of losses in the field as well as in storage, they become inadequate. Generally, about 30 % fruits and vegetables are rendered unfit for consumption due to spoilage after harvesting.

Post-harvest loss of fruits and vegetables has been defined as "that weight of wholesome edible product (exclusive of moisture content) that is normally consumed by human and that has been separated from the medium and sites of its immediate growth and production by deliberate human action with the intention of using it for human feeding but which for any reasons fails to be consumed by human." Not only quantity and quality but even the appearance of fruits and vegetables are affected and their market value is reduced. Most 'skin deep' injuries such as 'fly speck' in apples do not affect the edible part. Some infections are not harmful as they occur on inedible parts and can be trimmed off before.

Fresh fruits and vegetables are perishable and highly prone to these losses because they are composed of living tissues. These tissues must be kept alive and health throughout the process of marketing. These are composed of thousands of living cells which require care and maintenance. Therefore, the reduction of post-harvest loss of fruit and vegetables is a complementary means for increasing production. It may not be necessary to considerably step up the production of fruits and vegetables with the growing demand if the post-harvest loss is reduced to a great extent. The cost of preventing losses after harvest in general is less than preventing a similar additional amount of fruit and vegetable crop of the same quality. Attention to the concept of post-harvest food loss reduction, as a significant means to increase food availability, was drawn by the World Food Conference held in Rome in 1974.

The global dairy industry is impressive by large. In 2005, world milk production was estimated at 644 million tons, of which 541 million tons was cows' milk. The leading producers of milk were the European Union at 142 million tons, India at 88 million tons, the United States at 80 million tons (20.9 billion gallons), and Russia at 31 million tons. Cheese production amounted to 8.6 million tons in Western Europe and 4.8 million tons in the United States. The vast array of products made from milk worldwide leads to an equally impressive array of spoilage microorganisms. A survey of dairy product consumption revealed that 6% of US consumers would eat more dairy products if they stayed fresher longer. Products range from those that are readily spoiled by microorganisms to those that are shelf stable for many months, and the spoilage rate can be influenced by factors such as moisture content, pH, processing parameters, and temperature of storage.

Meat and Poultry Spoilage

Meat is a nutritious, protein-rich food which is highly perishable and has a short shelf-life unless preservation methods are used. Shelf life and maintenance of the meat quality are influenced by a number of interrelated factors including holding temperature, which can result in detrimental changes in the quality attributes of meat. Spoilage by microbial growth is the most important factor in relation to the keeping quality of meat. In most developing countries, including Nigeria, fresh meat forms a significant proportion of meat intake. It is either eaten cooked or processed into other forms to avoid associated spoilage. The main causative factor of such spoilage has been linked to unavailability of necessary storage facilities and favorable ambient temperature that usually prevail in developing countries that are in tropical regions Research findings have suggested that there is increasing attention on the use of naturally occurring metabolites produced by selected lactic acid bacteria (LAB) to inhibit the growth of spoilage microorganisms.

Spoilage and Preservation of Meat

Factors Affecting Meat Spoilage

Meat has long been considered a highly desirable and nutritious food, but unfortunately it is also highly perishable because it provides the nutrients needed to support the growth of many types of microorganisms. Due to its unique biological and chemical nature, meat undergoes progressive deterioration from the time of slaughter until consumption. In general, the metabolic activity of the ephemeral microbial association which prevails in a meat ecosystem under certain aerobic conditions, or generally introduced during processing, leads to the manifestation of changes or spoilage of meat. These changes or spoilage are related to the:

- Composition and population of the microbial association.
- The type and the availability of energy substrates in meat.

Indeed the type and the extent of spoilage is governed by the availability of low-molecular weight compounds (e.g., glucose, lactate) existing in meat. By the end of the phase changes and subsequently, overt spoilage is due to catabolism of nitrogenous compounds and amino acids as well as secondary metabolic reactions.

The post-mortem glycolysis, caused by indigenous enzymes, ceases after the death of the animal when the ultimate pH reaches a value of 5.4–5.5. Afterwards, the contribution of meat indigenous enzymes in its spoilage is negligible compared to the microbial action of the microbial flora. A number of interrelated factors influence the shelf life and keeping quality of meat, specifically holding temperature, atmospheric oxygen (O_2), indigenous enzymes, moisture (dehydration), light and, most importantly, micro-organisms. All of these factors, either alone or in combination, can result in detrimental changes in the colour, odour, texture and flavour of meat. Fresh meat has a shelf life of 1 day or less at ambient storage temperatures, 20-30 °C. Spoilage is said to be a state of a particular food in which it is offensive to consumers' senses, usually caused by metabolites of contaminant microorganisms. Meat spoilage is not always evident and consumers would agree that gross discoloration, strong off-odours, and the development of slime would constitute the main qualitative criteria for meat rejection. In general, spoilage is a subjective judgment by the consumer, which may be influenced by cultural and economic considerations and background as well as by the sensory acuity of the individual and the intensity of the change. Spoilage of meat can be considered as an ecological phenomenon that encompasses the changes of the available substrata, such as low molecular weight compounds, du ring the proliferation of bacteria that constitute the microbial association of the stored meat. The prevailing of a particular microbial community of meat depends on the factors that persist during processing, transportation and storage in the market. Such may vary widely from one country to another as a result of differences in climatic conditions, coupled with possible varying levels in knowledge of food hygiene practices of the handlers.

The microbiological quality of meat depends on the physiological status of the animal at slaughter, the spread of contamination during slaughter and processing, the temperature and other conditions of storage and distribution. In fact, some of the microorganisms originate from the animal's intestinal tract as well as from the environment with which the animal had contact at some time before or during slaughter. Other organisms, including psychrotrophic bacteria, are recovered from hides and work surfaces within an abattoir as well as from carcasses and butchered meat at all stages of processing.

A wide range of micro-organisms coming from different sources are introduced onto carcass surfaces, which contain abundant nutrients and which have high water availability. Only a few of the contaminants will be able to initiate growth, and only some of these will eventually spoil the meat by means of their biochemical attributes. Predominance of different groups of microorganisms on meat depends on the characteristics of the meat, the environment in which meat is stored as well as the processing that meat may undergo.

A vast number of studies in meat microbiology have established that spoilage is caused by only a fraction of the initial microbial association that comes to dominate. The range of microbial taxa found on meat is given in Table. A consortium of bacteria, commonly dominated by Pseudomonas spp., is in most cases responsible for spoilage of meat stored aerobically at different temperatures (-1 to 25 °C); the Pseudomonas spp. can grow under refrigeration temperatures. It is established that under aerobic storage three species of Pseudomonas, Ps. fragi, Ps. fluorescens and Ps. Lundensis, are the most important spoilage organisms. The population of pseudomonads to the level of 10^{7-8} CFU/g, has been attributed to slime and off-odours formation. However, in practice both these characteristics become evident when the pseudomonads have exhausted the glucose and lactate present in meat and begin to metabolise nitrogenous compounds such as amino acids. This is significant in dry firm dark meat where there is no carbohydrate and therefore spoilage occurs earlier at lower populations (10^6). Brochothrix thermosphacta and lactic acid bacteria (LAB) have been detected in the aerobic spoilage flora of chilled meat. These organisms have been isolated from beef carcasses during boning, dressing and chilling. Moreover, lairage slurry, cattle hair, rumen contents, walls of slaughter houses, the hands of workers, air in the chill room, neck and skin of the animal as well as the cut muscle surfaces have been shown to be contaminated with these organisms. Both LAB and Br. Thermosphacta is the main, if not the most important, cause of spoilage, which can be recognized as souring rather than putrefaction. Br. thermosphacta has been reported to be responsible for spoilage of meat products under refrigeration conditions.

Table: Genera of spoilage bacteria commonly found on meats and poultry.

Microorganisms	Gram reaction	Fresh	Processed
Achromobacter	-	X	
Acinetobacter	-	XX	X
Aeromonas	-	XX	X
Alcaligenes	-	X	

Bacillus	+	X	X
Brochothrix	+	X	X
Campylobacter	-	X	
Carnobacterium	+	X	
Chromobacterium		X	
Citrobacter		X	
Clostridium	+	X	
Corynebactenum	+	X	X
Enterobacter		X	X
Enterococcus	+	XX	X
Escherichia	-	X	
Flavobacterium	-	X	
Hafnia	-	X	X
Janthinobacterium	-		X
Klebsiella	-	X	
Lactobacillus	+	X	XX
Lactococcus	+	X	
Leuconostoc	+	X	X
Listeria	+	X	X
Microbacterium	+	X	X
Micrococcus	+	X	X
Moraxella	-	XX	
Proteus	-	X	
Providencia	-	X	X
Pseudomonas	-	XX	X
Shewanella	-	X	X
Staphylococcus	+	X	X
Streptococcus	+	X	X
Weissella	+	X	X
Yersinia	-	X	
X, known to occur; XX, most frequently isolated.			

Table: Common defects in meat products and causal bacteria.

Defect	Meat product	Bacteria
Slime	Meats	Pseudomonas, Lactobacillus, Enterococcus, Weissella, Brochothrix
H_2O_2 greening	Meats	Weisella, Leuconostoc, Enterococcus, Lactobacillus
H_2S greening	Vacuum packaged meat	Shewanella
H_2S production	Cured meats	Vibrio, Enterobacteriaceae
Sulfide odour	Vacuum Packaged meat	Clostridium, Hafnia
Cabbage odour	Bacon	Providencia
Putrefaction	Ham	Enterobacteriaceae, Proteus
Bone taint	Whole meats	Clostridium, Enterococcus
Souring	Ham	Lactic acid bacteria, Enterococcus, Micrococcus, Bacillus, Clostridium

Table: Factors and precursors affecting the production of odour end-products of Gram-negative bacteria, such as Pseudomonas spp., Shewanella putrefaciens and Moraxella.

End product	Factors	Precursors
Sulfur compounds Sulfides	Temperature and substrate (glucose) limitation	Cysteine, cystine, methionine
Dimethylsulfide		Methanethiol, methionine
Dimethyldisulfite		Methionine
Methyl mercaptan		nad
Methanethiol		Methionine
Hydrogen sulfide	High Ph	Cystine, cysteine
Dimethyltrisulfide	nad[a]	Methionine, methanothiol
Esters Methyl esters	Glucose (l)[b]	nad
Ethyl esters	Glucose (l)	nad
Aldehydes		
2-Methylbutanal	nad	iso-Leucine
Alcohols		
Methanol	nad	nad
Ethanol	nad	nad
2-Methylpropanol	nad	Valine
2-Methylbutanol	nad	iso-Leucine
Other compounds		
Ammonia	Glucose (l)	Amino acids

[a] nad, no available data.
[b] (l) low concentration of glucose.

Need and forms of Meat Preservation in Developing Countries

Owning to the spoilage potential of meat, many varieties of preservation techniques are employed in improving its keeping quality and shelf life. In good hygienic conditions, after slaughter and evisceration, the optimal way to preserve meat is under refrigeration at temperatures around 4 °C. Most African countries, because of lack of refrigeration facilities in the slaughter house, ambient temperatures above 20 °C and lack of suitable transportation between the production and marketing areas, meat can be exposed to conditions of high risk with respect to increased contamination resulting from growth of pathogens and spoilage microorganism. Although most regulations recommend meat to be kept under refrigeration, the fact is that in many areas of most developing countries this does not occur.

The majority of meat produced in abattoirs is sold for immediate consumption through retailers who buy from butchers and resell to consumers who usually subject it to cooking and consume within days. However, for various reasons, there are left-overs that are not sold. Since proper storage facilities are lacking, the left-over meat is processed into various forms in order to avoid spoilage. This involves improvising traditional techniques of preservation. In cases where modern storage methods (such as refrigerators and freezers), are available, they are either expensive to maintain or means for their maintenance (electricity) are lacking. As an alternative, meat is preserved by processing to semi-dry and dry forms. A typical example is kundi, a dry meat product produced by cutting raw meat into pieces which is then parboiled and sundried in an open container, made of materials that can conduct heat. This method of preparation makes the meat product prone to microbial and other sources of contamination. However, the product, being an intermediate moisture meat (IMM), is low in moisture content and is shelf stable under tropical climates without refrigeration.

Another product that meat is processed into is Suya. This is a popular, traditionally processed meat which is served or sold along streets, in club houses, on picnics and in restaurants. There are three main forms of suya, namely tsire, kilishi and balangu, but of these, tsire is the most commonly preferred. Therefore, to most consumers, tsire is synonymous with suya. Tsire is a roasted, boneless meat of beef, goat or mutton that is cooked around a glowing charcoal fire in which the meat pieces are staked on wood sticks, spiced with peanut cake, spices, vegetable oil, salt or other flavourings. It is a delicatessen item since it does not receive any treatments designed to extend its shelf life. Indeed, most sales-points hardly exhaust their sales and leftovers are often carried over to the second day or beyond. To this extent, rancidity often sets in, leading to the spoilage of this product. Suya products can become contaminated microbiologically from raw materials, handlers and/or equipment. Igene and Abulu reported the isolation of Bacillus, Streptococcus, Staphylococcus, Escherichia, Proteus, Pseudomonas and Klebsiella from raw and freshly roasted tsire subjected to different storage treatments. Researchers also reported the confirmation of some of these organisms in the stick meat, specifically Ps. aeruginosa, B. cereus, Staph. aureus, and E. coli.

Bacillus cereus is one of the organisms that could cause food borne disease associated with consumption of contaminated meat. The genus, Bacillus, is also known to cause souring in meat, while B. anthracis can cause disease in man, though regarded as a relatively low risk from meat and meat products. B. cereus is a ubiquitous organism and has been found in raw beef and milk, and the organism is directly linked to dairy cows. Therefore, contamination of carcasses of dairy cows is possible but is not thought to constitute a significant risk in foods of animal origin. Foodborne illness caused by B. cereus generally results from improper handling of foods. Proteus spp. have been found in small numbers in the flora on beef and pork carcases and in a variety of ready toeat processed meats. They have also been associated with the spoilage of beef; Proteus is known to be associated with putrefaction of meat.

Meat or meat products are not thought to be a major source of Staph. aureus as causative agent of food borne disease in man, even though the organism is an important pathogen in animals. The principal source of transmission between animals and man is unpasteurized milk and cheese made from un-pasteurized milk. Outbreaks of staphylococcal food poisoning in man are frequently associated with improper food handling and temperature abuse of foods of animal origin, but it is generally believed that the main source of contamination is food handlers. Nevertheless, strains of Staph. aureus can become endemic in food processing plants and meat can be contaminated from animal or human sources. Staph. aureus has been isolated from cattle carcasses and is also found in raw beef. The organism can become a major problem in cured meats as it is very salt tolerant and grows well when other flora are removed by the preservation methods.

Preservation using Lactic Acid Bacteria

A General Overview

The lactic acid bacteria comprise a group of Gram positive, non-sporulating, cocci or rods, and are catalaselacking organisms. LAB produces lactic acid as the major end product during the fermentation of carbohydrates. They only grow in complex media where fermentable carbohydrates and higher alcohols are used as an energy source, mainly to form lactic acid. Homofermentative LAB degrade hexoses to lactate, whereas heterofermentative LAB degrade hexoses to lactate and additional products such as acetate, ethanol, CO_2, formate, or succinate. LAB are widespread in most ecosystems and are found in soil, water, plants, and animals. They are responsible for many food fermentation processes, but they are also commonly found on non-fermented foods such as dairy products, meat products, seafood, fruits, vegetables, cereals, sewage, and in the genital, intestinal, and respiratory tracts of humans and animals. LAB are widely used as protective cultures in the food industry for the production of fermented foods, including dairy (yogurt, cheese), meat (sausages), fish, cereals (bread and beverages such as beer), fruit (malolactic fermentation processes in wine production), and vegetables (sauerkraut, kimchi, silage).

Most LAB are considered as 'generally recognized as safe', GRAS. They are used to ensure safety, preserve food quality, develop characteristic new flavours, and improve the nutritional qualities of food. LAB exert strong antagonistic activity against many related and unrelated microorganisms, including food spoilage organisms and pathogenic bacteria such as Listeria, Clostridium, Staphylococcus and Bacillus spp. The antagonistic effect of LAB is mainly due to a lowering of the pH of the food, to competition for nutrients, and to the production of inhibitory metabolites. LAB are able to grow at refrigeration temperatures. They tolerate modified atmosphere packaging, low pH, high salt concentrations, and the presence of additives such as lactic acid, ethanol, or acetic acid.

The classification of LAB is based on morphological, metabolic and physiological criteria. As described earlier, LAB are related by a number of typical metabolic and physiological features. In the past few decades, DNA based methods targeting genes such as 16S rRNA, applied to determine the relatedness of food-associated LAB, have resulted in significant changes in their taxonomic classification. The genera comprising LAB are Lactobacillus, Leuconostoc, Pediococcus, Lactococcus, and Streptococcus, as well as Aerococcus, Carnobacterium, Enterococcus, Oenococcus, Teragenococcus, Vagococcus, and Weisella. Members of the LAB typically have a G+C content below 50%.

LAB in Meat

In meats, LAB constitute a part of the initial microflora which develops easily after meat is processed to fermented sausages, chill stored or packed under vacuum or modified atmosphere. The strains of LAB generally considered as being found naturally in meats and meat products are: Carnobacterium piscicola and C. divergens; Lactobacillus sakei, Lb. curvatus and Lb. plantarum; Leuconostoc mesenteroides subsp. mesenteroides, Leuc. gelidum and Leuc. carnosum. LAB in fresh meat bring about a mild fermentation process without producing any changes in the sensory characteristics because of the low carbohydrate content and the strong buffering capacity of meat. In the same way the growth of LAB in naturally fermented meats, after the addition of sugar, transforms the products through the production of lactic acid by the LAB. The subsequent decrease in pH denatures the meat proteins favouring the decrease of water activity (aw), which ends up in a microbial stabilization of the transformed product.

In addition to the fermentable carbohydrates, glucose, glycogen, glucose-6-phosphate and small amounts of ribose, meat and meat products provide a number of vital growth factors such as available amino acids and vitamins that support the growth of the fastidious LAB. Some species of the genera Lactobacillus, Carnobacterium, Leuconostoc spp. and Weissella are especially well adapted to this ecosystem. Several representatives of the genus Lactobacillus may typically dominate the microbial population especially of vacuum packaged and processed meat products. The facultative heteroferemtative species of Lb. sake and Lb. curvatus are found in most meat systems and are probably the most frequently encountered species of the genus. These two species have been shown to be of major economic importance in meat products, or acting as main and desirable fermentative organisms in dry sausages.

Persistence and competitive ability of Lactobacillus and several other species of the genera Leuconostoc (Leuc. amelibiosum, Leuc. carnosum, Leuc. gelidum), Weissella (W. viridescens, W. halotolerans) and Carnobacterium (Cb. divergens, Cb. piscicola) in processed meat systems are explained by their ability to ferment the carbohydrates in meat and their adaptation to the meat substrate. While the leuconostocs appear to grow most rapidly on chilled fresh meat, Lb. curvatus and Lb. sake, on account of their higher tolerance of elevated salt concentrations and nitrite, typically dominate raw fermented sausage and pasteurized emulsified meat products. Some of these features also apply to two species of the genus Pediococcus, Ped. pentosaceus, and Ped. acidilactici, which are associated with fermented meat products.

Enterococcus and Lactococcus are other genera of LAB that are of some commercial significance. Enterococcus spp. use the homolactic pathway for energy production, yielding mainly L(+) lactic acid from glucose at pH values less than 5. At pH values above 7, ethanol, acetic acid and formic acid are the main products of glucose fermentation. In the absence of heme and under aerobic conditions, glucose is converted to acetic acid, acetoin and carbon dioxide. The genus Enterococcus differs from the lactococci by their resistance to 40% bile and growth of most species at 6.5% salt. E. faecium and E. faecalis are associated with the gastro-intestinal tract of man and warm-blooded animals and have been suggested as indicators of faecal contamination of meat.

The association of pediococci with proteinaceous foods such as fresh and cured meat, and raw sausages, has frequently been reported and particularly for Ped. acidilactici and Ped. pentosaceus in fermented sausages. The association of pediococci with meat fermentations has been a topic of intensive study. Meat and meat products provide a favourable growth substrate for strains of Ped. acidilactici and Ped. pentosaceus, and particularly in the fermentation of semidry sausages or other cured products, such strains appear to play some role during fermentation and maturation. Pediococci are also frequently found in vacuum or modified-atmosphere packaged meat and meat products, in which the LAB population is, however, most often dominated by species of the genera Lactobacillus, Leuconostoc, Carnobacterium, Weissella and Enterococcus.

Use of LAB as Biological Preservatives

Biopreservation, preservation by the use of biological agents, refers to the extension of the shelf-life and improvement of the safety of foods using microorganisms and/or their metabolites. Antagonistic cultures which are added to meat products to inhibit pathogens and/or prolong the shelf life, while changing the sensory properties as little as possible, are termed protective cultures. Their antagonism refers to inhibition through competition for nutrients and/or production of one or more antimicrobially active metabolites. In a recent study by Olaoye and Onilude, the potential of selected species of Pediococcus as biological preservatives in the extension of shelf life of fresh beef was investigated. The authors reported that the LAB strains used were able to effect preservation of the meat product, for few days before spoilage was started to set in. In a similar study, Olaoye and Dodd also reported the extension in shelf life of tsire, a traditional Nigerian stick meat, after treatment with bacteriocinogeic cultures of Pediococcus.

Nowadays, the consumer pays a lot of attention to the relation between food and health. As a consequence, the market for foods with health-promoting properties, so called functional foods, has shown a remarkable growth over the past few years. Also, the use of food additives is regarded as unnatural and unsafe. Yet, additives are needed to preserve food products from spoilage and to improve the organoleptic properties; hence the use of functional protective cultures in the food fermentation industry is being explored. Functional protective cultures are microorganisms that possess at least one inherent functional property. The latter can contribute to food safety and/or offer one or more organoleptic, technological, nutritional, or health advantages. The implementation of carefully selected strains as microbial cultures or co-cultures in fermentation processes can help to achieve in situ expression of the desired property, maintaining a perfectly natural and healthy product. Examples are LAB that are able to produce antimicrobial substances, sugar polymers, sweeteners, aromatic compounds, useful enzymes, or nutraceuticals, or LAB with health-promoting properties, so called probiotic strains. This represents a way of replacing chemical additives by natural compounds, at the same time providing the consumer with new, attractive food products and it also leads to a wider application area and higher flexibility of cultures.

Production of Antimicrobials by LAB for Food Preservation

In meat, production of one or more antagonistic metabolites may be part of the complex mechanism by which a micro-organism becomes established in the presence of other competing organisms. The understanding of such mechanisms provides a valuable key to our understanding the complexity of microbial interactions in a meat system and hence the basis of 'biological' approaches to food preservation. One of the main roles of LAB in biopreservation is to improve safety by inactivating pathogens and spoilage microorganisms via acid production and bacteriocins. Furthermore, it is essential that potential biopreservative cultures show no pathogenic or toxic activities. The food industry is expected to produce safe, healthy and nutritious products of high quality. For many food products, fermentation with starter cultures containing lactic acid bacteria (LAB) is an essential part of the production process.

Organic Acid Production

An important role of meat LAB starter cultures is the rapid production of organic acids; this inhibits the growth of unwanted flora and enhances product safety and shelf life. The antimicrobial effect of organic acids lies in the reduction of pH, and in the action of undissociated acid molecules. It has been proposed that low external pH causes acidification of the cytoplasm. The lipophilic nature of the undissociated acid allows it to diffuse across the cell membrane collapsing the electrochemical proton gradient. Alternatively, cell membrane permeability may be affected, disrupting substrate transport systems. The types and levels of organic acids produced during the fermentation process depend on the LAB strains present, the culture composition, and the growth conditions.

Fermentation of the carbohydrates, glucose, glycogen, glucose-6-phosphate and small amounts of ribose, in meat and meat products, produces organic acids by glycolysis or the Hexose Monophosphate, HMP pathway. L (+) lactic acid is more inhibitory than its D(-) counterpart. Lactic acid is a major fermentation end product of LAB and a number of other genera (e.g Brochothrix). The LAB in particular are able to reduce the pH to levels where putrefactive (e.g. clostridia and pseudomonads), pathogenic (e.g. salmonellas and Listeria spp.) and toxinogenic bacteria (Staphylococcus aureus. Bacillus cereus, Clostridium botulinum) will be either inhibited or killed. Also, the undissociated acid, on account of its fat solubility, will diffuse into the bacterial cell, thereby reducing the intracellular pH and slowing down metabolic activities, and in the case of Enterobacteriaceae such as E. coli inhibiting growth at around pH 5.1. The rapid reduction of the pH below 5.3 during sausage fermentation is sufficient to inhibit growth of salmonellas and Staph. aureus.

Table: Metabolic products of lactic acid bacteria with antimicrobial properties.

Product	Main target organisms
Organic acids	
Lactic acid	Putrefactive and Gram-negative bacteria, some fungi.
Acetic acid	Putrefactive bacteria, clostridia. some yeasts and fungi.
Hydrogen peroxide	Pathogens and spoilage organisms. especially in protein-rich foods.
Low-molecular-weight metabolites	
Reuterin (3-OH-propionaldehyde)	Wide spectrum of bacteria, moulds and yeasts.
Diacetyl	Gram-negative bacteria.
Fatty acids	A range of different bacteria.
Bacteriocins Nisin	Some LAB and Gram-positive bacteria, notably endospore-formers.
Others	Gram-positive bacteria, inhibitory spectrum according to producer strain and bacteriocin type.

Table: Classification of bacteriocins from lactic acid bacteria.

Category	Subcategory
Class I—lantibiotics	Type A: elongated molecules. Subtype A1: leader peptides are cleaved by a dedicated serin proteinase. Subtype A2: leader peptides are cleaved by a dedicated ABCtransporter. Type B: globular molecules
Class II—nonmodified, heat-stable bacteriocins	Class IIa: Pediocin-like bacteriocins. Class IIb: Two-peptide bacteriocins. Class IIc: Sec-dependent bacteriocins. Class IId: Other bacteriocins.
Class III—large, heat-labile bacteriocins	

Bacteriocin Production by LAB

The bacteriocins of LAB possess common traits that justify their classification on a sound scientific basis into three well defined classes:

- Class I, the lantibiotics, small heat-stable polycyclic peptides (<5 kDa) containing small, membrane active peptides.

- Class II, the small (<10 kDa) heat-stable non-lantibiotics such as pediocin-like bacteriocins with a strong anti-Listeria activity.
- Class III, large (>30 kDa) heat-labile bacteriocins.

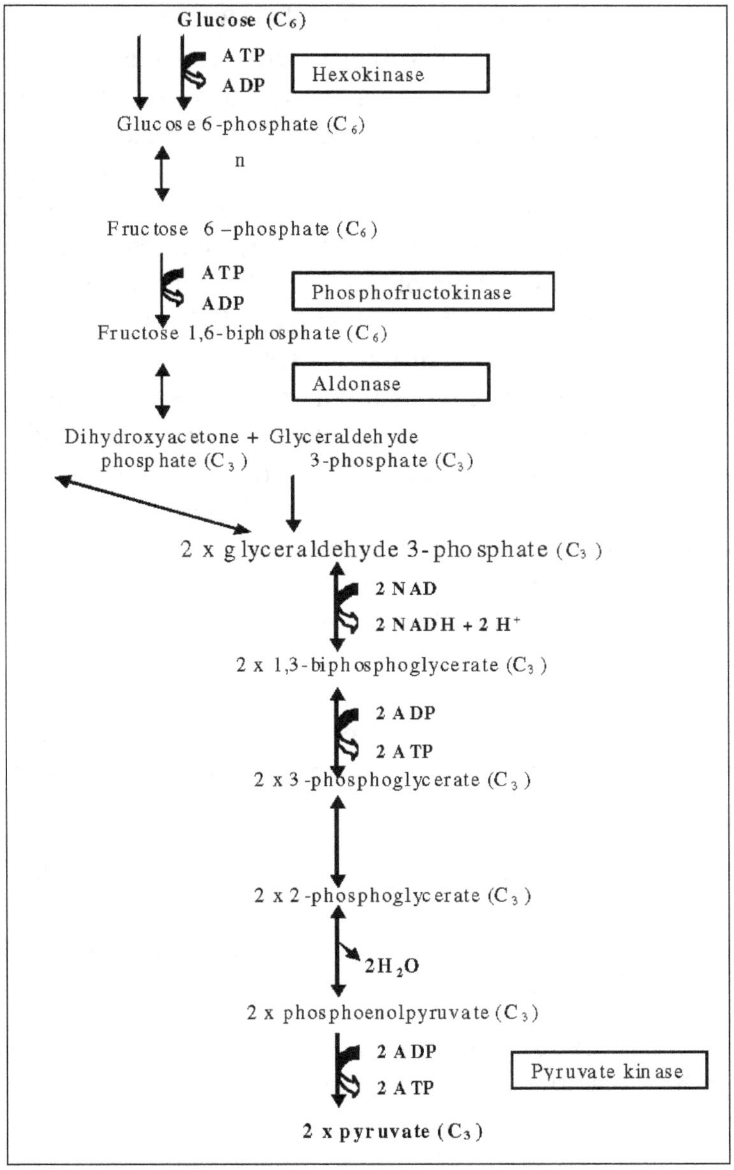

Embden Meyerhof Parnas pathway.

Due to their abundance and possible application in industrial processes, bacteriocins belonging to the first two classes are the most thoroughly studied. The most prominent Class I bacteriocin is nisin, which is produced by strains of Lactococcus lactis subsp. lactis isolated from milk and vegetable-based products and by Lc. lactis BB24 isolated from Spanish-dry fermented sausages. Nisin is a broad spectrum bacteriocin with bactericidal activity towards a wide range of Gram-positive bacteria, including Staphylococcus aureus and Lis.monocytogenes. In addition, nisin prevents spore outgrowth and inhibits vegetative cells of Bacillus spp. and Clostridium spp. To date, nisin is the most thoroughly studied and characterized bacteriocin of LAB and the only one internationally accepted as a food biopreservative in certain foods.

Class II bacteriocins (non-lantibiotics) comprise a heterogeneous group of bacteriocins. Despite differences in their primary structures, most Class II bacteriocins are small (<10 kDa) and heat-stable peptides with a high content of small amino acids such as glycine. They are usually cationic and often amphiphilic, reflecting their ability to kill target cells by permeabilizing the cell membrane. Class IIa bacteriocins are the most thoroughly studied LAB bacteriocins and possess interesting technological properties and a strong antimicrobial activity against a broad range of Gram-positive spoilage and food-borne pathogens, especially Lis. monocytogenes. The search for LAB producing antilisterial bacteriocins has lead to the description and characterization of a large number of Class IIa bacteriocins, produced by a wide variety of Pediococcus, Leuconostoc, Enterococcus, Lactobacillus and Carnobacterium strains. Class IIa is also referred to as the pediocin family, which is named from pediocin PA-1, the first and most thoroughly characterized bacteriocins within the group. Pediocin-like bacteriocins, members of the class II bacteriocins, are of considerable commercial interest owing to their characteristics of being small, heat resistant peptides that are not modified posttranslationally. All the pediocin-like bacteriocins share certain features, including a seven amino acid conserved region in the N-terminal of the active peptide. They are active against other LAB but are particularly effective against Lis. monocytogenes. Pediocin PA-1 is, perhaps, the best known, produced by Pediococcus acidilactici isolated from American-style sausages and Ped. pentosaceus Z102 from Spanish style sausages. In the past, several pediocin PA-1-producing LAB strains were independently isolated in different laboratories. However, in many cases the bacteriocin produced received different names (pediocins PA-1, AcH, JD, Bac and 347, mesentericin 5) before identification and realization that all were the same molecule. The pediocin PA-1-containing fermentate AltaTM 2341 is a commercial food ingredient reported to extend the shelf life of a variety of foods and, particularly, to inhibit the growth of Lis. monocytogenes in ready-to-eat meat products. The determination of the pediocin PA-1 amino acid sequence, the application of improved protocols for its purification, and the identification of the pediocin PA-1 operon have been reported.

The high molecular weight Class III bacteriocins have been identified within the genera Lactobacillus and Enterococcus. These bacteriocins, in contrast to Class I and II bacteriocins, are inactivated upon heat treatment (e.g., 60 100 °C for 10 15 min) and, similar to type B lantibiotics, do not act on sensitive cells by membrane-disruption.

Interest in the bacteriocins produced by meat LAB has increased dramatically, reflecting their growing importance with respect to the functional properties of starter cultures. A number of bacteriocins are produced by most LAB species involved in meat fermentation, including Lb. sakei, Lb. curvatus, Lb. plantarum, and Ped. acidilactici. Meat-borne LAB produce a range of bacteriocins that are generally active towards other LAB (contributing to the competitiveness of the producing strain) and food borne Gram–positive pathogens such as Lis. monocytogenes, Staph. aureus, C. perfringens and B. cereus. Bacteriocins exert their inhibitory action via the formation of pores in the cytoplasmic membrane of sensitive cells as well as interrupting DNA and protein syntheses. Generally, bacteriocins target the cell envelope and, with the exception of the larger proteins (>20 kDa) that degrade the murein layer (e.g. lysins and muramidases), use non-enzymatic mechanisms to cause the depolarization of the target cell membrane and/or inhibit cell wall synthesis. Bacteriocins have generally a cationic character and easily interact with Gram-positive bacteria that have a high content of anionic lipids in the membrane determining the formation of pores. Pores in the cytoplasmic membrane clearly affect the energetic status of the cell, i.e. dissipation of proton motive force (PMF) causing an arrest of ΔpH and Δψ (trans-membrane electrical potential) dependent processes (such as transport) while certain bacteriocins cause ATP efflux. A bacteriocin producer protects itself against its own antimicrobial compound by means of a system referred to as immunity, which is expressed concomitantly with the antimicrobial peptide. The mode of action of bacteriocins can be bactericidal or bacteriostatic, determining death or extension of lag phase respectively. In Gram-positive bacteria, the bacteriocin nisin produced by Lc. lactis has been shown to act on energized membrane vesicles to disrupt PMF, inhibit uptake of amino acids, and cause release of accumulated amino acids. Studies on the mode of action of bacteriocins have indicated that bactericidal activity was confined to pH values of 6 and lower. This is possibly due to the influence of two positively-charged (lysine) and two negatively-charged (glutamate and aspartate) amino acids and two histidine residues with a positive charge at pH 6 or lower (pK_a = 6 for His) and having a major role in determining the effective charge of the peptide which is crucial for activity. Gram-negative bacteria are protected by their outer membrane, which prevents bacteriocins from reaching the plasma membrane.

It is generally accepted that bacteriocin activity is less effective in meat products than in in vitro systems. Activity may be reduced by the binding of the bacteriocin molecules to food components (mainly the fat matrix), and by the destabilizing action of proteases and other enzymes. Further limitations of bacteriocin effectiveness are uneven distribution in the food matrix and their inhibition by salt and curing agents. Even so, several authors report that certain bacteriocinogenic meat LAB could be used as bio-protective cultures to prevent the growth of pathogens in sausage. Indeed, the use of bacteriocin-producing Lactobacillus sakei as a starter culture decreases the numbers of Listeria in fermented sausage. Antilisterial effects have also been demonstrated with bacteriocinogenic Lb. curvatus, Lb. plantarum and Ped. Acidilactici. The production of bacteriocins with a broad inhibition range, especially towards food-borne pathogens is therefore highly desirable since this would ensure the competitiveness of the starter strain while reducing the numbers of harmful flora.

Other Antimicrobials of LAB

Hydrogen peroxide is produced from lactate by LAB in the presence of oxygen as a result of the action of flavoprotein oxidases or nicotinamide adenine dinucleotide (NADH) peroxidises. The antimicrobial effect of H_2O_2 may result from the oxidation of sulfhydryl groups causing denaturing of a number of enzymes, and from the peroxidation of membrane lipids thus increasing membrane permeability. H_2O_2 may also be a precursor for the production of bactericidal free radicals such as superoxide (O^{-2}) and hydroxyl (OH^-) radicals which can damage DNA. The enzyme catalase hydrolyses hydrogen peroxide. Some LAB strains involved in meat fermentation, such as Lb. sakei, Lb. plantarum, Lb. pentosus and Ped. acidilactici, possess heme-dependent catalase activity which is active in meat products since these substrates contain heamin in abundance. Most undesirable bacteria such as Pseudomonas spp. and Staph. aureus are many times more sensitive than the LAB to H^2O^2.

Carbon dioxide is mainly produced by hetero-fermentative LAB. The precise mechanism of its antimicrobial action is still unknown. However, CO_2 may play a role in creating an anaerobic environment which inhibits enzymatic decarboxylation, and the accumulation of CO_2 in the membrane lipid bilayer may cause a dysfunction in permeability. CO_2 can effectively inhibit the growth of many food spoilage microorganisms, especially Gram-negative psychrotrophic bacteria. The degree of inhibition by CO_2 varies considerably between the organisms. CO_2 at 10% (v/v) could lower the total bacterial counts by 50% (v/v) and at 20–50% it had a strong antifungal activity. Pathogens (e.g Enterobacteriaceae and Listeria) could also be inhibited due to reduced pH effects as CO_2 dissolves to produce a weak acid.

Diacetyl, an aroma component, is produced by strains within all genera of LAB by citrate fermentation. It is produced by hetero-fermentative lactic acid bacteria as a by-product along with lactate as the main product. Diacetyl is a high value product and is extensively used in the dairy industry as a preferred flavour compound. Lb. rhamnosus gives a high yield for diacetyl, 64 mg of diacetyl per g of glucose consumed. The physiological reason for the production of diacetyl is not clearly understood. It is hypothesized that diacetyl is synthesized to reduce the toxicity of pyruvate. Diacetyl also has antimicrobial properties. It inhibits the growth of Gram-negative bacteria by reacting with arginine utilization. Jay showed that Gram-negative bacteria were more sensitive to diacetyl than Gram-positive bacteria; the former were inhibited by diacetyl at 200μg/ml. The antimicrobial activity of diacetyl was evaluated against E. coli, Lis. monocytogenes and Staph. aureus in a study by Researchers the authors concluded that the organisms were sensitive to diacetyl with Lis. monocytogenes having the least susceptibility. Generally varying concentrations of diacetyl are required to bring about inhibitions of different pathogenic and spoilage organism.

Beneficial Effects of LAB on Meat

Strains of LAB to be used in the biopreservation must be carefully selected in order to achieve the desired beneficial effect. This is because not all LAB cultures can be used to achieve the purpose. The use of LAB as biological preservatives on meat products could confer health benefits to the consumers.

Seafood Spoilage

It is generally agreed that the flora of fish found in temperate waters consist of Gramnegative psychrotolerant bacteria, whose growth is possible at 0 °C but optimal around 25 °C. Among these, the majority belong to the subclass γ of proteobacteria: Pseudomonas, Shewanella, Acinetobacter, Aeromonas, Vibrio, Moraxella, Psychrobacter, Photobacterium, etc. and to a lesser extent the CFB (CytophagaFlavobacter-Bacteroides) group. Nevertheless, Gram-positive bacteria like Micrococcus, Bacillus, Lactobacillus, Clostridium or Coryneforms, may also be present in variable proportions. Some genera, like Vibrio, Photobacterium and Shewanella, require the presence of salt to multiply and are thus typically found in seawater while Aeromonas is more common in fresh water even though it is often isolated from marine products. In tropical fish, the flora has the same composition overall, but often with a greater proportion of Gram-positive bacteria (Micrococcus, Bacillus, Coryneforms) and enterobacteria.

The indigenous microflora of the gastro-intestinal tract of fish have been much more studied than those of the skin or the mucus due to their importance in digestion, nutrition and growth and in disease control in aquaculture. Although this environment is partially anaerobic, most researchers have observed a predominance of aerobic bacteria, also present in the surrounding water, capable of surviving and multiplying in the particular medium of the gastro-intestinal tract. This predominance of aerobic bacteria could be due to the collecting techniques, which are not always suitable for strict anaerobes. Nevertheless, researchers have shown by molecular methods that the aerobic flora of rainbow trout intestine usually represents 50 90% of the total flora. Gram-negative bacteria dominate the intestinal flora. In general, Aeromonas, Pseudomonas and members of the Flavobacterium/Cytophaga group are most often found in the intestine of freshwater fish while Vibrio, Acinetobacter, and Enterobacteriaceae are more common in marine fish. These are fermentative bacteria that develop rapidly in the gastrointestinal tract due to the low pH, the lack of oxygen and the abundance of nutrients. Staphylococci have also been found to be the dominant flora in the intestine of the Arctic char. Although not predominant, lactic acid bacteria (Lactobacillus, Carnobacterium, Streptococcus, Leuconostoc, Lactococcus, Vagococcus) have often been isolated from the gastro-intestinal tract of fish.

Both the number and diversity of the microflora are probably widely underestimated because the majority of studies carried out so far have used classic microbiological methods involving growth on agar media. The cultivability of bacteria has been estimated to be sometimes less than 2% of intestinal flora rainbow trout, or even less than 0.01% of the skin flora. They have identified a strict anaerobe (Clostridium gasigenes) in the intestinal flora of rainbow trout. Similarly, researchers have shown the presence of Clostridium in the intestinal mucus. Moreover, molecular methods have enabled a new species belonging to the genus Mycoplasma to be detected for the first time in fish. It was found in abundance in the intestine of wild and farmed salmon.

The worldwide shrimp market is mainly composed of the Nordic shrimp (Pandalus borealis), which is only fished, and the tropical shrimp (Penaeus sp.), which can be fished or farmed and whose production has expanded rapidly in recent years. The deep-water tropical shrimp (Parapenaeus longirostris) is also found in Europe, particularly on the Spanish and Portuguese markets.

As in fish, the bacterial flora of shrimps depends on several factors including the species considered, the geographic location and environment, the temperature and salinity of the water, etc. However, overall, the same species of microorganisms are found in shrimps and in fish from a given geographical zone. In fresh tropical shrimps, the initial bacterial flora consists mainly of Pseudomonas, Vibrio, Acinetobacter, Moraxella, Flavobacterium and a high proportion of Aeromonas. In India, researchers have detected significant amounts of different species of Vibrio, including V. parahaemolyticus. Researchers working on Nicaraguan shrimps, reported a predominance of Coryneforms and Moraxella followed by lower levels of Bacillus, Lactobacillus, Micrococcus, Proteus, Shewanella, Acinetobacter and Pseudomonas. These results confirm those previously obtained by Matches. Researchers have shown the influence of the fishing zone on the nature of the initial flora: mostly Gram-positive bacteria on shrimps fished at low depths and Pseudomonas on those caught in deep water. The nature of this initial flora has an effect on the shelf life and the organoleptic properties of shrimps linked to spoilage.

Microbial Contamination of Flesh and Its Evolution During Preservation

At fish death, the immune system collapses and bacteria can contaminate the flesh by moving through the muscle fibres. However, bacteria are found in much greater quantities on the skin than in the tissues and it is more likely that the spoilage of the whole fish is mostly due to bacterial enzymes that have spread through the tissues. On the other hand, the various stages of processing (evisceration, head removal, filleting, and trimming) contribute to spreading the bacteria naturally present in fish throughout the muscular tissue, thus accelerating spoilage.

Regarding the bacterial growth, first, there is a lag phase, generally short for fish found in temperate waters but whose length varies depending on the composition and state of the fish, the storage temperature and the species of bacteria. Next, the microorganisms begin a period of exponential growth and can reach level of 10^{6-8} CFU g^{-1}. Clearly, the rate of multiplication depends on the factors mentioned previously. Generally, tropical fish kept in ice have a longer lag phase (1 to 2 weeks) than that of fish found in temperate waters and growth during the exponential phase is slower, probably because the microorganisms cannot adapt well to low storage temperatures.

The Concept of Specific Spoilage Flora

Sensory spoilage is not always linked to the total number of microorganisms. In most seafood products, organoleptic rejection occurs well after the total flora has reached its maximum. Although the flora of fish is varied, only certain microorganisms, named "specific spoilage microorganisms" are responsible for the production of unpleasant odours and flavours.

In general, in unprocessed fish, the chemical modifications that lead to sensory rejection are due to only one bacterial species. In the case of processed products, however, the mechanisms are often more complex as several bacterial groups can interact and contribute to the spoilage of the product. The specific spoilage flora and associated chemical molecules vary according to the species of fish, the storage temperature, the type of packaging or processing and even the fishing season.

Identifying the microorganisms specifically responsible for spoilage is not easy. One method consists in analysing the products during storage (selective microbiological enumeration, sensory and chemical analyses) and establishing correlations between different parameters. However, the enumerations are not always selective enough and it is often more interesting to collect colonies from the total flora at the time of sensory spoilage and then check their capacity to produce bad odours in a pure culture. Tests done directly in the fish matrix always give more relevant results than using a laboratory medium or fish flesh extract but are often more difficult to carry out as the fish must be sterilised without modifying its composition. The collections of fillets in aseptic conditions or low intensity ionisation are the solutions most frequently used. Once the microorganisms potentially involved in spoilage have been identified, it is important to check their colonisation kinetics on naturally contaminated products and during "real" storage conditions (temperature, packaging). Potential spoilage flora in a model medium may not have the capacity to develop enough to reach levels that could cause perceptible sensory deterioration. Furthermore, interactions with endogenous flora could modify the spoiling characteristics of microorganisms.

Bacterial Metabolism

Fish is a matrix that particularly favours microbial development. Despite a low percentage of carbohydrates (0.2 to 1.5% depending on the species), fish flesh is rich in non-protein, low molecular weight nitrogenous molecules that are rapidly metabolised by bacteria. These compounds include free amino acids, creatine, nucleotides, urea and trimethylamine oxide (TMAO). The high post-mortem pH (>6) and the low acidification during preservation, combined with the small quantity of carbohydrates present, enable the rapid growth of pH-sensitive psychrotrophic spoilage bacteria like Shewanella putrefaciens. Lastly, fatty fish are rich in polyunsaturated fatty acids that can be rapidly oxidised by either chemical chain reactions or lipolysis resulting from autolytic or bacterial enzyme activity.

In aerobiosis, the carbohydrates (ribose and lactate) can be metabolised into CO_2 and H_2O. In anaerobiosis, and in the presence of an electron acceptor such as TMAO, some microorganisms are capable of anaerobic respiration that leads to the production of acetic acid. However, spoilage of fresh fish is rarely linked to the production of organic acids due to the low concentration of carbohydrates in flesh. TMAO, whose concentration varies according to the species, plays an extremely important role in spoilage as certain microorganisms such as Shewanella, Photobacterium and Aeromonas reduced it to trymethyl amine (TMA), a pungent molecule responsible for the strong amine odour typical of rotten fish. TMAO is occasionally found in freshwater fish but is generally associated with marine fish.

Urea found in large quantities in selachians can be metabolised into ammonia which also has a strong, unpleasant odour. Deamination of amino acids also leads to the production of ammonia. The breakdown of sulphurous amino acids found naturally in fish leads to the production of hydrogen sulphide (H_2S) (from cysteine), methylmercaptan and dimethyl-disulphide (from methionine). These molecules all play a part in the spoilage process. Certain microorganisms such as Shewanella, Photobacterium and lactic acid bacteria are capable of this production, but to an extent that varies depending on the strain.

The decarboxylation of amino acids leads to the formation of biogenic amines, which are often linked to spoilage even though they have no particular odour in the product. Tyrosine is a precursor of tyramine and cadaverine. Arginine can be degraded into putrescine via the agmatine pathway in the presence of arginine decarboxylase, as is the case with Photobacterium. Putrescine can also be produced by the decarboxylation of ornithine, for example in the case of certain enterobacteria such as Hafnia alvei and Serratia liquefaciens. Histamine is a biogenic amine formed by the degradation of histidine. It can provoke allergic-like reactions of varying intensity (redness of the skin, swelling, headaches) and fish with high levels of histidine are strictly regulated. In fish, the principal producers of histidine are mesophilic enterobacteria such as Morganella morganii, Hafnia alvei and Klebsiella pneumoniae, but recently psychrotolerant microorganisms such as Photobacterium phosphoreum and Morganella psychrotolerans have been clearly incriminated in cases of histamine poisoning.

Spoilage of Fresh Fish and Shellfish

Fish

Fish spoilage occurs at varying speeds depending on a large number of parameters. Different fish species display a variety of surface characteristics in terms of the resistance of their skin texture and the composition of the mucus, which may contain antibodies and bacteriolytic enzymes. Geographical zone, in particular water temperature, also plays an important role as fish from tropical waters keep much longer in ice than those from temperate or cold waters. Rough handling of the fish can damage its integrity and accelerate spoilage. The storage conditions (temperature and packaging) also play a vital part in fish preservation.

When temperate-water sea fish (like cod for example) are stored at chilled temperature, Sh. putrefaciens are the specific spoilage bacteria. Research has shown that the number of bacterial cells is inversely correlated to the remaining shelf life, and consider that at approximately 10^8 CFU g^{-1} the sensory deterioration of the product is no longer acceptable. The spoilage is characterised by the production of TMA, H_2S and other sulphurous compounds such as methylsulphide and dimethylsulphide. Sh. putrefaciens consists of a heterogeneous group whose taxonomy has greatly changed over the last few years. Researchers have shown that the large majority of microorganisms that produce H_2S isolated from Baltic Sea fish chilled for several days should now be identified not as Sh. putrefaciens, but as Sh. baltica. Furthermore, other minor strains have been identified as new species: Sh. hafniensis, Sh. morhuae, Sh. glacialipiscicola and Sh. Algidipiscicola. Tropical sea or freshwater fish are spoiled by Pseudomonas. Sh. putrefaciens has also been isolated from these products but does not appear to play an important part in spoilage. This could be due to the inability of this microorganism to develop in the presence of a large number of Pseudomonas. The spoilage caused by this microorganism can be differentiated from Sh. putrefaciens by the absence of TMA and sulphurous compounds and the appearance of fruity and rotten odours caused by aldehydes, ketones and esters. When fish is stored at room temperature, Aeromonas is more likely to spoil freshwater fish stored in aerobiosis, but it has also been shown that Sh. putrefaciens can be involved.

The Effects of Storage Temperature on Spoilage and Shelf-life

Storage temperature is the most influential factor on shelf life. Researchers have described models that express the relation between relative spoilage speed and storage temperature. The relative spoilage speed at a temperature T is defined as the ratio of the shelf life at 0 °C to the shelf life at T °C. For example, if the shelf life of a cod is 14 days at 0 °C and 6 days at 5 °C, the relative speed of spoilage will be equal to 14/6=2.3, i.e. the spoilage will be 2.3 times faster at 5 °C than at 0 °C. Shelf life at 0 °C differs depending on the species of fish and the method of preservation, but the effect of temperature on the relative spoilage speed R is constant and the following formula has been established: $\sqrt{R} = 1+0.1 * T$ °C. This "square-root" model enables the shelf life of a product to be calculated at different temperatures if its shelf life is known at a certain temperature.

The bacteria responsible for the spoilage of fresh fish are different depending on the storage temperature. Sh. putrefaciens, Pseudomonas sp., Aeromonas sp. and Ph. phosphoreum are the principal spoilage bacteria found between 0 and 5 °C, and their quantity varies depending on the storage atmosphere. At 15-30 °C, Enterobacteriaceae, Vibrionaceae and Gram-positive bacteria are responsible for spoilage. The "square-root" model described previously does not take into account these changes in flora. Nevertheless, the estimations of relative spoilage speed are satisfactory for whole fresh fish, vacuum-packed fish or modified atmosphere packaged (MAP) fish. However, for tropical fish, the relative spoilage speeds at 20-30 °C are more than double those estimated by the model. A separate model for tropical fish has therefore been developed for temperatures between 0 and 30 °C. It establishes a linear correlation between the Neperian logarithm of the relative spoilage speed and the storage temperature (Ln R = 0.12* T °C).

Super-chilling consists in storing products at temperatures between 0 and 4 °C. The "square-root" model gives satisfactory results for products stored in this way. This technique can greatly extend the shelf life of the product but often lower the product quality (water retention, texture, etc.).

The Effects of Storage Atmosphere on Shelf-life: Anaerobiosis and CO_2

Vacuum-packaging or MAP with varying amounts of CO_2 (25-100%) are both widely used to preserve food. Numerous studies have been carried out on seafood products with very different results. Most frequently, high concentrations of CO_2 lead to a 30 to 60% longer shelf life. The efficiency is closely linked to the temperature, which must be as low as possible so that the gas can dissolve in the product. However, unlike in meat products, these techniques do not significantly lengthen fish shelf life. In fact, although the number of Pseudomonas decreases due to the lack of oxygen in vacuum-packed temperate-water sea fish, Sh. putrefaciens can respire in anaerobiosis thanks to TMAO, and can develop up to 10^6 -10^8 CFU/g, thus increasing the production of TMA. Nevertheless, it is probable that, below 10^8 CFU g^{-1}, Sh. putrefaciens is not the only microorganism responsible for fish spoilage as these authors have also identified the presence of Ph. phosphoreum on vacuum-packed cod. This microorganism, which had until recently escaped detection by microbiologists produces 10 to 100 times more TMA per cell than Sh. putrefaciens. Consequently, these two microorganisms can provoke the spoilage of vacuum-packed temperate water sea fish and it is probably the initial quantity of each type that determines which one will dominate the flora. Furthermore, the growth rate of Ph. phosphoreum increases in anaerobiosis, which explains why this microorganism plays such an important part in the spoilage of packaged products such as cod.

In the presence of CO_2, the growth of Sh. putrefaciens and Pseudomonas is greatly inhibited whereas Ph. phosphoreum is relatively resistant. It reduces TMAO to TMA without producing H_2S, which explains why MAP products are characterized by high levels of TMA without the odour of H_2S typical of spoiled fish stored in aerobiosis. Ph. phosphoreum can be eliminated if the raw material is first frozen, which increases the shelf life of thawed cod stored in a modified atmosphere at 2 °C from 12 to 20 days.

Numerous other seafood products have a shelf life similar to that of cod. As Ph. phosphoreum is widespread in the marine environment, it seems likely that this microorganism, or others resistant to CO_2, are responsible for the spoilage of packed seafood products). Indeed, researchers have shown that Ph. phosphoreum is the specific spoilage bacteria of MAP salmon. Researchers have shown that Atlantic halibut (Hippoglossus hippoglossus) packed in a CO_2/O_2 (50/50) enriched atmosphere has a microflora dominated by Ph. phosphoreum and, due to the presence of oxygen, Pseudomonas sp. as well as Brochothrix thermosphacta on samples at the end of their shelf life. Ph. phosphoreum has also been found on pollock and tuna from the Indian Ocean (Thunnus albacares).

Ph. phosphoreum constitutes a heterogenic group whose taxonomy has evolved over the last few years. Some strains isolated from seafood products have characteristics that are not homogenous, notably in luminous and non-luminous strains. For this reason, some of the strains isolated from cod fillets in a modified atmosphere have been reclassified as Ph. Iliopiscarium.

The efficiency of MAP has also been demonstrated for warm-water fish have identified Br. thermosphacta as the dominant species on sea bream packed in a CO_2-enriched atmosphere (40%) at the end of their shelf life. This microorganism could play a part in spoilage. Researchers have demonstrated that, for tropical tuna steaks, a CO_2/O_2 (40/60) modified atmosphere is better than vacuum-packing for inhibiting the growth and histamine production of the psychrotrophic Morganella.

The combination of super-chilling and modified atmosphere enabled salmon fillets to retain a high level of quality after 24 days of storage due to the almost total inhibition of psychrotrophic bacterial growth. The shelf life is 3 to 4 times longer than that of a refrigerated product under air.

Shellfish

Amongst shellfish, Nordic and tropical shrimps are by far the most consumed products worldwide and this section will be wholly devoted to them. Tropical shrimps (Penaeus sp.) are generally frozen shortly after catch and sold frozen before being processed (usually cooked) on land. Nordic shrimps (Pandalus borealis) are generally cooked after catch, sometimes peeled, and then frozen.

The few studies concerning this product have shown that shrimps are very sensitive to spoilage. This sensitivity is due to a relatively high pH (>7) and the fact that the intestinal tract containing enzymes and bacterial flora is not removed immediately after catch. In addition, the muscle contains large quantities of non-protein, water-soluble molecules such as amino acids. These amino acids are used directly by the bacteria for their growth. Shrimps generally contain large amounts of arginine and glycine and small amounts of cysteine and methionine.

Shelf life is variable and depends on the nature of the initial flora and the spoilage bacteria, often unknown, that may develop during storage. It can be relatively short: 5 to 8 days for ice-packed Penaeus merguiensis; 6 days at 0 °C for Parapenaeus longirostris. Other authors have found longer shelf lives for tropical shrimps: 13 to 16 days when ice-packed; 10 to 17 days when icepacked or over 20 days when stored in liquid ice for different species of the genus Penaeus found in Australian waters.

The major cause of early shrimp spoilage is melanosis, which is the formation of black spots on the cephalothorax. This process is biochemical and is not due to bacterial activity, but rather an enzyme complex called polyphenol oxidase. The benzoquinones thus produced interact with amines, amino acids and oxygen to form the melanin responsible for this black coloration.

Other organoleptic changes are generated by the action of bacteria. Researchers have shown that when tropical Australian shrimps are icepacked in an oxygen-rich environment, the growth of Pseudomonas fragi increases, whereas using liquid ice where oxygen is limited leads to the development of Sh. putrefaciens. The sensory characteristics of contaminated products are very different depending on the predominance of one or the other of these bacteria. To confirm these observations, researchers have demonstrated experimentally that strains of Ps. fragi inoculated into shrimp juice produce odours of fruit and onion, whereas strains of Sh. putrefaciens give off a sulphurous odour. Moreover, these two bacteria display different profiles of volatile compounds.

ining 4045% CO_2 with 30 or 5% O_2, respectively, delays microbial growth and the production of TVBN and TMA compared to air-packed or iced stored shrimp, especially at the end of storage. Amine production and the low level of H_2S-producing microorganisms and enterobacteria suggest that P. phosphoreum could be involved in the spoilage of shrimps both in air and MAP. The production of biogenic amines (tyramine, putrescine, cadaverine, agmatine) during storage is higher for atmospheres with modified O_2 concentrations than for air. The same species of shrimp and using the same gas compositions, shelf life can be prolonged at least by 2 days (9 days instead of 4 to7 days with the ice storage). This type of packaging can also extend the shelf life of cooked whole shrimps (Pandalus borealis) by 200% compared to air-packed products.

Spoilage Flora of Lightly Preserved Products

Semi-preserved seafood products have undergone a very light preservation process, such as salting, drying, smoking or marinating, leading to a final pH greater than 5 and a salt content below 6% in the aqueous phase. Examples include smoked fish, carpaccio of salmon or trout, gravelax, marinated anchovy fillets, cooked products like peeled shrimps, preserved in brine or in a modified atmosphere etc. In terms of microbiology, these products are very fragile. Because the raw material follows a long preservation process with a good deal of handling, there is a significant risk of recontamination. Moreover, no step of the process involves the total elimination of microorganisms and the final product is often preserved for several weeks at low temperature before being eaten. This allows some psychrotrophic bacteria, which can spoil organoleptic qualities or present a risk to human health, to multiply. We shall deal with two examples representative of this product category, cold smoked salmon and cooked peeled shrimps, due to their great economic importance in the European market.

Cold-Smoked Salmon

In France, there is a long tradition of salting, drying and cold-smoking seafood products, mainly salmon and herring. Nowadays, in the industrialised countries, the aim of smoking is not so much to ensure a long shelf life for the product but to give it a particular colour and taste. With more than 20,000 tonnes produced each year, France has always been the world leader in the production and consumption of smoked salmon. It was originally a luxury product, eaten only on special occasions and made by traditional methods. In recent decades, however, smoked salmon has become an everyday food, available in supermarkets all year round and produced on an industrial scale.

The production of smoked salmon involves the different steps of filleting, salting, drying and smoking at low temperatures (< 25 °C) but each plant has developed its own technology: salting by dry salt or by injection, smoking with different wood essences and different smoke generators etc. The final salt concentration is usually between 2.5 and 3.5% in the flesh while the smoke content, estimated by the quantity of total phenols, is between 2 and 20 ppm. These products are generally sold sliced and vacuum-packed. The producer is responsible for the use-by date, which is usually three to six weeks at 4 °C in Europe. However, it is not uncommon to notice deterioration in taste from the end of the second week of preservation.

It has been clearly shown that this organoleptic spoilage is mainly due to microbial activity, with autolytic or chemical reactions of lipid oxidation playing a lesser role. For a very long time, however, the mechanisms of microbial spoilage of smoked fish were poorly understood. Although lactic flora have always been found to be dominant in this product, their role in spoilage was not clear because no correlation between their number, nor even that of the total flora, and the sensory deterioration could be identified. More recent studies have specified the composition of the flora of these products and their link with spoilage.

On leaving the factory, product contamination can range from 10^2 to 10^6 bacteria/g, the initial level varying mainly according to the factory and its hygiene standards and independently of the origin and state (fresh or frozen) of the raw material. Despite the presence of inhibitory factors like salt, smoke, preservation at low temperature and vacuum packaging, bacterial growth in these products can be quite rapid. It is not unusual to observe total flora concentrations of 10^6 CFU/g from the first week of storage, which can regularly reach 10^{7-9} CFU/g before the use-by date.

The initial flora of smoked salmon is often dominated by Gram-negative bacteria typical of fresh fish flora, such as Shewanella, Photobacterium, and Aeromonas, later identified as Serratia. Nevertheless, as well as Photobacterium, researchers have also found Brochothrix, Yersinia and

Carnobacterium. During vacuum-packed preservation at low temperatures, the diversity of genera decreases and Gram-positive bacteria, especially lactic acid bacteria, very often become predominant. Variations between factories and even between batches from the same factory have often been observed but the lactic flora always seems to dominate, regardless of the geographical provenance of the products analysed (processed in Europe). 50 to 90% of the colonies taken from the total flora of samples at the end of preservation are lactic acid bacteria. The technological parameters of salting and smoking can influence the final proportion of this group of bacteria so it should be possible to control the level of lactic flora by varying these parameters.

In smoked salmon, the major species are Carnobacterium maltaromaticum and Lactobacillus curvatus or Lb. sakei. Other species such as Carnobacterium divergens, Lactobacillus farciminis, Lb. alimentarius, Lb. plantarum, Lb.homohiochii, delbrueckii, Lb. casei, Lb. coryneformys, Leuconostoc mesenteroides, Enterococcus faecalis and Weisella kandleri are more rarely isolated. Despite this predominance, quite often the number of enterobacteria (Ph. phosphoreum and Br. thermosphacta) is also fairly high. In contrast, yeasts are rarely present, except sometimes when the product contains a high concentration of salt or phenol. Nevertheless, they do not reach high enough population levels to contribute to spoilage. To illustrate the complexity of the flora of smoked salmon, researchers have shown that of 96 isolates coming from 30 batches of smoked salmon in Spain and preserved for 3 weeks at 2 °C, 49% were lactic acid bacteria, 20% enterobacteria, 16% micrococci (mostly coagulase-negative staphylococci), 5% Br. thermosphacta, 5% Gram-negative (Moraxella, Acinetobacter and Pseudomonas), 2,5% mobile Aeromonadaceae and 2,5% Bacillus.

Recently, temporal temperature or denaturating gradient gel electrophoresis (TTGE, DGGE) molecular methods have been applied to smoked salmon. In some cases, these techniques have given similar results to the culture method (predominance of Lactobacilli, Ph. phosphoreum and Ph. iliopiscarium at the end of preservation. In contrast, only found clones corresponding to Gram-negative bacteria (Vibrio, Photobacterium, Enterobacteriaceae, Alteromonas) and assumed that this technique is biased because it does not allow a good amplification of the DNA of Gram-positive bacteria. Nevertheless, in both studies, clones no doubt corresponding to new species of Photobacterium and Vibrio were identified by the culture-independent technique, underlining how these tools can complement classic culture methods.

The role of lactic acid bacteria in the sensory deterioration of smoked salmon is still not very clear. Several authors have shown that there is no correlation between the total lactic flora and sensory spoilage. However, were able to prolong the shelf life of smoked salmon by inhibiting lactic acid bacteria with nisine (bacteriocin), suggesting a possible spoilage effect by this bacterial group. By inoculating cubes of smoked salmon sterilised by ionisation, certain species of lactic acid bacteria were very spoiling (i.e. Lb. sakei) while others had no effect at all (i.e. Lb. alimentarius). However, the potential for spoilage seems to vary according to the strain tested. Lb. sakei generally produces sulphurous and acidic odours, associated with the production of H_2S, acetic acid, ethyl and n-propyl-acetate, but some Lb. sakei do not spoil the organoleptic qualities of this product. Similarly, Lb. alimentarius, which does not spoil smoked salmon, has been identified as the bacteria responsible for the sensory deterioration of marinated herrings. The role of Carnobacteria is still under discussion. Many studies have shown that they are probably not very spoiling in smoked salmon. Inoculation by different strains of Cb. maltaromaticum and Cb. divergens results in no or few changes in organoleptic qualities. When the Carnobacteria reach a sufficient level, odours of butter and plastic may be detected, associated with the production of 2, 3-butanedione (diacetyl) and

2,3-pentanedione, but these odours/flavours do not lead the product to be rejected by a specialist jury and thus even less so by a jury of consumers.

Among the other bacteria frequently found in smoked salmon, Researchers. showed that Se. liquefaciens was very spoiling, releasing odours of amines, cheese, acid or rubber, associated with the molecules TMA, dimethyldisulphur, 2,3 butanediol and 2-pentanol. However, Se. liquefaciens is considered less spoiling than Lb. sakei as the unpleasant odours are perceived much later. Br. thermosphacta also leads to the sensory rejection of the product due to the odours of blue cheese and plastic, well correlated with 2heptanone and 2-hexanone. Nevertheless, it is quite rare for these bacteria to reach sufficiently high levels in naturally contaminated products to be the sole explanation for sensory rejection. Although strongly spoiling in fresh fish packed in a modified atmosphere, Ph. phosphoreum seems to play a more moderate role in the deterioration of smoked salmon. Weak odours of "acid", "amine" and "feet" result in the product being judged as moderately spoiled. Moreover, there is a great variability according to the strain. Researchers give much greater weight to the spoiling action of this species. They have shown a good correlation between the sensory quality of smoked salmon and the production of tyramine and histamine, two potential chemical indicators for spoilage. Sh. putrefaciens, the most common spoilage bacteria in fresh fish, and Vibrio sp. have never been implicated in spoilage phenomena in smoked salmon, even when inoculated at high concentrations.

Although the bacteria responsible for sensory deterioration are now quite well identified, spoilage remains a complex phenomenon because the interactions between all these bacterial groups change their metabolism. For example, when Lb. sakei is coinoculated with Se. liquefaciens, spoilage is clearly delayed. Conversely, when Cb. maltaromaticum and Vibrio sp. are both present (two non-spoiling bacteria in pure culture) spoilage is increased significantly. Cb. divergens did not produce TVBN in pure culture in sterile smoked salmon but some TVBN was detected when the bacterium was added to commercial samples containing a natural endogenous flora. For example, a high production of putrescine (>200 μg/g) cannot be due to the simple degradation of ornithine, whose concentration does not exceed 10 μg/g in salmon flesh. Arginine deaminase-positive lactic acid bacteria like Cb. divergens and Lb. sakei, metabolise arginine, found in greater quantity in salmon, into ornithine, which is then converted to putrescine by ornithine decarboxylase-positive enterobacteria (Se. liquefaciens and Hafnia alvei).

All these results show that it is over-optimistic to expect to predict the quality of lightly preserved seafood products using only one microbiological or biochemical parameter. Not only is the microflora very complex but also the organoleptic properties of spoilage are very varied, with odours that can be described as amine, sulphur, acidic, sour or "cabbage". Some authors have succeeded in improving the organoleptic quality of smoked salmon by modifying the technological parameters, like the salt, the smoke or the addition of bio-protective bacteria, but have not demonstrated the inhibition of a particular group of bacteria. The situation is not the same in fresh fish, preserved in air or in a modified atmosphere or vacuum packed, where one or two well identified species dominate the flora and for which a predictive model of sensory spoilage has been suggested. However, in the case of smoked salmon, a multiple approach has enabled some authors to correlate quality to several microbiological and biochemical measurements. For example, a predictive model for the remaining shelf life of smoked salmon, based on TVBN content and enumeration of the flora on Rogosa medium at pH 5.5 (quite selective for lactobacilli). If the TVBN content is less than 35 mg N/100 g of flesh, the product is not spoiled. If it rises above this level, the number of bacteria on Rogosa medium must be considered. A product with 50 mg N/100 g of TVBN will only be spoiled if the number of colonies on Rogosa is higher than 10^5 CFU/g. Another model, based on pH and the quantity histamine and tyramine, has been developed by researchers.

Cooked Shrimps Packed in a Modified Atmosphere

France is the second biggest importer of shrimps in Europe, and this product corresponds to the highest value of all seafood imported. Tropical shrimps account for 80% of the shrimps imported. Refrigerated cooked ready-to-eat shrimps represent the largest shrimp consumption in France. They may or may not be peeled, are often packed under a CO_2-enriched protective atmosphere, and can also pass through brine containing salt and different organic acids (citric, ascorbic, benzoic, etc.). Thanks to increasing consumption, the shrimp market has been steadily expanding over the last ten years.

In the literature to date, knowledge of this product's microbiology has focused on the Nordic shrimp (Pandalus borealis). The flora of fresh shrimps is greatly reduced during the cooking stage of processing (70–80 °C throughout) and the spoilage flora is seemingly the result of a recontamination by Gram-positive bacteria before or during packaging. The microorganisms are then selected according to the different characteristics of the product and the storage conditions. More specifically, the shelf life and development of the microflora depend on the initial contamination, the characteristics of the product (brined or not for example) and the storage conditions (modified atmosphere, temperature). The combined effect of these different parameters could create a barrier effect against microbial development.

Brined Shrimps

The spoilage microflora was dominated at 5 °C by yeasts and lactic acid bacteria (Streptococcus sp., Lactobacillus sp.). Another study of brined Nordic shrimps by Einarsson and Lauzon demonstrated the importance of benzoate and sorbate in preventing the development of flora at 4.5 °C. When the flora is fully developed, it is dominated by coryneform bacteria and Moraxella sp.

Researchers have studied Nordic and tropical shrimps (Penaeus sp.) that had been brined, drained, packed in a modified atmosphere and stored at different temperatures between 0 and 25 °C. They showed that the effect of temperature on shelf life is even greater than with other seafood products. For example, batches at 8 °C had a shelf life 15 to 33 times longer than those stored at 25 °C. Certain variations in the composition of the product, no matter how small, can have a large effect on this shelf life. Thus, shelf life at 5 °C is over three months for tropical shrimps containing 2.3% salt in an aqueous solution and over 6 months for Nordic shrimps with 3.3% salt. The initial pH of these products is between 5.7 and 5.9. TVBN levels are relatively low when the products are spoiled (20 mg/100g at 15 and 25 °C and 10 mg/100g at the end of shelf life for tropical shrimps stored at lower temperatures). Most of the spoilage flora has not been identified, but lactic acid bacteria make up an important part of the total microflora, whatever the storage temperature. Amongst these bacteria, several strains isolated from products stored at 15 and 25 °C have been identified as Enterococcus faecalis. This microorganism is probably responsible for spoilage at these temperatures, but does not appear to develop much, if at all, at temperatures between 0 and 8 °C. The authors also observed the presence of Lb. curvatus on tropical shrimps stored at 0 °C.

In continuing their studies on the flora isolated from these products, observed an evolution of the microflora from bacilli to cocci between 5 and 8 °C for Nordic shrimps and between 8 and 15 °C for tropical shrimps. At 25 °C, the flora is dominated by cocci. Most of the strains isolated at different temperatures have been identified as Carnobacterium (Cb. divergens) and Enterococcus (En. faecalis). En. faecalis is present on products stored at 15 °C and higher whereas Cb. divergens and Lb. curvatus are found on products stored at 8 °C and lower. Three of the isolated strains correspond to an unknown Carnobacterium species. At high concentrations, En. faecalis and En. durans could represent a health risk for consumers. In this study, tyramine may represent a chemical indicator of spoilage, but is at levels low enough not to present a risk. The authors recommend storing the shrimp at low temperature and with at least 3% salt (water phase) so as to avoid spoilage problems and health risks.

Researchers have studied Nordic shrimps that have been either brined or brined and drained then packed in a modified atmosphere. They focused on the effect of hygiene (during industrial or manual processing), the composition of the brine and the storage conditions (atmosphere and temperature) on the overall evolution of the microflora and the shelf life. Different groups of organic acids were tested: benzoic, citric and ascorbic or acetic, citric and lactic, including the effect of diacetate. The pH of brined shrimps is between 5.6 and 5.8. The shelf life depends on the nature of the organic acids, their concentration, the temperature and also the initial contamination, which seems to have a significant effect. In fact, industrial products are more contaminated than products processed manually, which leads to shorter shelf lives and a more diverse spoilage flora. This is especially true for brined and drained MAP shrimps, which have a shelf life of over 75 days at 7 °C for manually processed products, but only 28-35 days for industrial products. Similarly, shrimps in brine composed of acetic, citric and lactic acid have shelf lives of 69-84 days and 42-49 days for manual and industrial products, respectively, at the same temperature. The modified atmosphere seems to prolong the shelf life of shrimps from 53-60 days for brined shrimps to over 75 days for the same shrimps that have been drained and packed in a modified atmosphere, although these results were not confirmed by subsequent tests.

The composition of the brine influences the nature of the flora that will develop on the shrimps. When it contains benzoic, citric and sorbic acid, the microflora is dominated by En. faecalis-like, Leuconostoc pseudomesenteroides, coagulase-negative Staphylococcus, yeasts, Ps. fluorescens, En. malodoratus, Cb. maltaromaticum and Lb. sakei. Adding diacetate to the brine can inhibit the growth of Ps. fluorescens, coagulase-negative Staphylococcus sp. and Enterococcus sp. When the brine contains acetic, citric and lactic acid, the microflora is composed exclusively of yeasts and lactic acid bacteria, such as Lb. sakei, which are responsible for unpleasant sour and buttery odours. When the yeasts dominate the flora, 10^6 CFU/g is enough to spoil the product.

Whatever the composition of the brine at the start of the process, the spoilage microflora of brined and drained and MAP shrimps are dominated by Lb. sakei. Spoilage of the industrial product leads to sour flavours and a release of gas in the packaging, probably produced by Lb. sakei. The spoilage of some brined shrimp samples with a reduced microflora is due to oxidation rather than microbial activity.

Shrimps Without Brine

Cooked, peeled, frozen or unfrozen Nordic shrimps without brine and packed in a modified atmosphere (50% CO_2) have an initial pH of 7.7, a salt content of 1.9% in aqueous solution and a TVBN of 10mg/100mg. If the product is frozen for 4 months preceding refrigerated storage then the initial concentration of microorganisms will be reduced, but the rate of growth, the maximum level of flora and the shelf life remain the same. The latter is 26 days at 2 °C, 16 days at 5 °C and 10 days at 8 °C. A low production of TVBN (lower than 20mg/100g) and no production of TMA, organic acids or biogenic amines were observed. From 116 isolates, 60% were identified as Cb. maltaromaticum, 27% were identified as Br. thermosphacta and 13% were identified as Psychrobacter sp. When inoculated together, Cb. maltaromaticum and Br. thermosphacta produced the same unpleasant odours as spoiled products that have been contaminated naturally. This is not the case when the two products are inoculated separately. The mechanisms of this interaction have not yet been explained. The presence of Br. thermosphacta has already been observed in MAP fish. Furthermore, this microorganism has already been described as producing odours of butter, blue cheese and feet. Cb. maltaromaticum produces an unpleasant chlorine odour whereas in previous studies this bacterium was described as non-spoiling only producing a buttery odour. Psychrobacter sp. does not seem to play a part in the spoilage of cooked shrimps.

The potential for spoilage by Carnobacterium on cooked, peeled, MAP Nordic shrimps without brine varies depending on the species and the strain. Cb. divergens and some Cb. maltaromaticum produce unpleasant odours whereas another group of Cb. maltaromaticum is not spoiling. Spoilage bacteria produce ammonia, tyramine, different alcohols, aldehydes and ketones. These authors have confirmed that Br. thermosphacta and Cb. maltaromaticum cultured together lead to some bad odours (wet dog) that are not produced when these species are grown separately. However, this association does not produce new metabolites and consequently the "wet dog" odour does not come from a metabiotic phenomenon in which a bacterium produces a metabolite from a substrate formed by another bacterium. On the other hand, Cb. maltaromaticum decreases the formation of diacetyl by Br. thermosphacta and the latter reduces the activity of Cb. maltaromaticum. The "wet dog" smell may be due to the interaction between metabolites formed by Cb. maltaromaticum, Cb. divergens, and partly by Cb. mobile, and those formed by Br. thermosphacta. The presence of oxygen increases the potential for spoilage and the number of metabolites produced by Br. thermosphacta. In fact, Br. thermosphacta can be aerobiotic or anaerobiotic depending on the respective percentages of O_2 and CO_2. To reduce the potential for spoilage in shrimps, it is therefore recommended to use an atmosphere with less O_2 and more CO_2.

In contrast to cooked, peeled, MAP shrimps without brine, in which the specific spoilage bacteria seem well defined, the characteristics of the spoilage flora of brined or brined/drained and MAP shrimps are not as clearly established. The composition of this spoilage flora depends on a large number of parameters: the initial contamination, the preservation parameters (organic acids and salt) and the storage conditions (atmosphere, temperature).

Fruit and Vegetable Spoilage

Consumption of fruit and vegetable products has dramatically increased in the United States by more than 30% during the past few decades. It is also estimated that about 20% of all fruits and vegetables produced is lost each year due to spoilage. The focus of this chapter is to provide a general background on microbiological spoilage of fruit and vegetable products that are organized in three categories: fresh whole fruits and vegetables, fresh-cut fruits and vegetables, and fermented or acidified vegetable products.

During the period 1970–2004, US per capita consumption of fruits and vegetables increased by 19.9%, to 694.3 pounds per capita per year. Fresh fruit and vegetable consumption increased by 25.8 and 32.6%, respectively, and far exceeded the increases observed for processed fruit and vegetable products. If US consumption patterns continue in this direction, total per capita consumption of fresh fruits and vegetables would surpass consumption of processed fruits and vegetables within the next decade.

This shift toward overall increased produce consumption can be attributed, at least in part, to increased awareness in healthy eating habits as revealed by a broad field of research addressing food consumption and health and promoted by the National Cancer Institute with the 5-A-Day Challenge along with the USDA-revised Food Pyramid. Additional factors influencing greater fresh produce consumption are the increased availability of fresh produce throughout the country throughout the year, increased diversity of selection at the retail level, and rapid growth in the fresh-cut, ready-to-eat produce sector (see "Microbiological Spoilage of Fresh-Cut Fruits and Vegetables").

According to a USDA-Economic Research Service study in 1995, 18.9 billion pounds of fresh fruits and vegetables were lost annually due to spoilage, which was 19.6% of all US losses of edible foods that year. The portion of loss specifically due to microbiological spoilage was not reported.

Most microorganisms that are initially observed on whole fruit or vegetable surfaces are soil inhabitants, members of a very large and diverse community of microbes that collectively are responsible for maintaining a dynamic ecological balance within most agricultural systems. Vectors for disseminating these microbes include soil particles, airborne spores, and irrigation water. Most bacteria and fungi that arrive on the developing crop plant either are completely benign to the crop's health or, in many instances, provide a natural biological barrier to infestation by the subset of microorganisms responsible for crop damage.

Spoilage microorganisms can be introduced to the crop on the seed itself, during crop growth in the field, during harvesting and postharvest handling, or during storage and distribution. Those same types of soil-borne spoilage microbes that occur on produce are the same spoilage microorganisms that are present on harvesting equipment, on handling equipment in the packinghouse, in the storage facility, and on food contact surfaces throughout the distribution chain. Therefore, early intervention measures during crop development and harvesting through the use of good agricultural practices (GAP) will provide dramatic reductions in yield loss due to spoilage at all subsequent steps in the food-to-fork continuum. Examples of GAPs include foliar fungicide application in the field, cross-contamination prevention measures (stringent sanitation standard operating procedures) in the packinghouse and storage facility, and use of postharvest fungicides. These practices also will enhance substantially the food safety and shelf life of fresh-cut produce (see "Microbiological Spoilage of Fresh-Cut Fruits and Vegetables").

In 1998, FDA published the Guide to Minimize Microbial Food Safety Hazards for Fresh Fruits and Vegetables, recommending GAPs that growers, packers, and shippers implement to address the common microbiological hazards that may be associated with their operations. These GAPs are organized in eight categories:

- Water
- Manure and municipal bio-solids
- Worker health and hygiene

- Sanitary facilities
- Field sanitation
- Packing facilities sanitation
- Transportation
- Traceback

In addition, FDA worked with the produce industry to develop commodity specific food safety guidelines for sprouts, lettuce and leafy greens, melons, and tomatoes that provided metrics for soil and water amendments as well as adjacent land usage.

Unusual Characteristics of Spoilage Microorganisms

Many fruits and vegetables present nearly ideal conditions for the survival and growth of many types of microorganisms. The internal tissues are nutrient rich and many, especially vegetables, have a pH near neutrality. Their structure is comprised mainly of the polysaccharides cellulose, hemicellulose, and pectin. The principal storage polymer is starch. Spoilage microorganisms exploit the host using extracellular lytic enzymes that degrade these polymers to release water and the plant's other intracellular constituents for use as nutrients for their growth. Fungi in particular produce an abundance of extracellular pectinases and hemicellulases that are important factors for fungal spoilage. Some spoilage microbes are capable of colonizing and creating lesions on healthy, undamaged plant tissue. Spoilage microorganisms also can enter plant tissues during fruit development, either through the calyx (flower end) or along the stem, or through various specialized water and gas exchange structures of leafy matter. Successful establishment, however, requires the spoilage microbe to overcome multiple natural protective barriers. Fruits and vegetables possess an outer protective epidermis, typically covered by a natural waxy cuticle layer containing the polymer cutin. A diverse community of epiphytic microorganisms that present a further competitive barrier to the spoilage organism also typically colonizes the outermost fruit surface. Overcoming these barriers requires an exquisite set of biochemical tools that allow the spoilage microorganism to (1) identify and recognize the plant surface; (2) employ one or more strategies to achieve irreversible attachment to the plant surface; and (3) initiate steps leading to internalization of the tissue. On plant structures other than the fruit, internalization can be achieved through a number of specialized vessels and surface structures employed by the plant to absorb and release water and to provide CO_2 and O_2 exchange.

Table: Bacterial fruit pathogens[a].

	2004 Annual US per capita consumption (lbs)[b]	Pseudomonas	Erwinia	Xanthomonas	Acidovorax
Apples	18.8		+		
Bananas	25.8				
Berries	6.1				
Citrus	22.7	+		+	
Grapes	7.9				
Melons	14.7				+

Peaches	5.1				
Pears	3.1		+		
Pineapple	4.4				

[a]Important postharvest diseases retrieved from researchers.
[b]85.2% of all fresh fruits consumed per capita in the United States in 2004.

However, the fruit of the plant lacks many of these structures, requiring the spoilage microbe to employ other methods to become internalized. This may partially explain the rather limited success of bacteria to spoil fruits and an improved ability to spoil vegetables that are not the fruit of the plant. The natural acidity of most fruits also serves as a barrier to many spoilage microbes, especially bacteria. By contrast, spoilage fungi that typically produce more diverse and greater amounts of extracellular depolymerases successfully attack and spoil both fruits and vegetables.

Colonization and lesion development more typically and more rapidly occurs within damaged or otherwise compromised plant tissue. External damage such as bruising, cracks, and punctures creates sites for establishment and outgrowth of the spoilage microbes. Lesion development can be relatively rapid, occurring within days or weeks. This presents the risk that rapidly reproducing spoilage microorganisms will arrive within open wound sites at the packing facility, and thereby, through shedding from the asymptomatic wound, present the potential for cross-contamination within the facility during handling, culling, washing, sorting, and packing before storage. Such cross-contamination to some degree is inevitable and, if not carefully managed with a robust facility sanitation program, could lead to the establishment of a population of spoilage microbes endemic to the facility that may be difficult to eradicate. A further and potentially more serious complication is the introduction into the cold storage facility of spoilage microorganisms already established in wound sites on product, whether the product is in bins or boxed and palletized. Depending upon storage conditions and storage time (greater than 12 months for certain robust crops), and if not carefully managed, these "primed" spoilage microorganisms can have a devastating impact on the stored product. Apples, for example, are stored in very large, controlled atmosphere storage rooms, either in wooden bins or boxed and ready for distribution.

Table: Fungal fruit pathogens[a].

	2004 Annual US per capita consumption (lbs)[b]	Penicillium	Geotrichum	Fusarium	Botrytis	Colletotrichum	Mucor	Monilinia	Rhizopus	Phtyophthora
Apples	18.8	+			+	+	+	+		
Bananas	25.8			+		+				
Berries	6.1	+			+	+	+	+		+
Citrus	22.7	+	+			+				+
Grapes	7.9	+			+				+	
Melons	14.7									
Peaches	5.1	+			+			+	+	
Pears	3.1	+			+		+			
Pineapple	4.4			+						

[a]Important postharvest diseases retrieved from researchers.
[b]85.2% of all fresh fruits consumed per capita in the United States in 2004.

Table: Bacterial vegetable pathogens[a].

	2004 Annual US per capita consumption (lbs)[b]	Pseudomonas	Erwinia	Xanthomonas	Bacillus	Clostridium	Lactic acid bacteria
Broccoli	5.9	+	+	+			
Cabbage	8.3	+	+	+			
Carrots	8.9	+	+		+		
Corn, sweet	9.6						
Cucumbers	6.3		+		+		
Lettuce, head	22.5	+	+	+			
Lettuce, leaf	12.0	+	+	+			
Mushrooms	2.6	+					
Onions	21.7		+		+		
Potatoes	46.5	+	+		+	+	
Spinach	2.1						
Tomatoes	19.3	+	+	+	+		+

[a]Important postharvest diseases retrieved from researchers
[b]81.0% of all fresh vegetables consumed per capita in the United States in 2004.

Extensive blue mold infestation on apples.

Two wound pathogens, Penicillium expansum and Botrytis cinerea, if not scrupulously cleaned from fruits prior to storage or if fruits with infected wounds have not thoroughly been culled from the lot, can cause significant crop loss as these spoilage fungi eventually degrade the wound sites, create lesions, and cross-contaminate adjacent fruits. If fruits receive improper preharvest fungicide application, poor washing, and/or inadequate culling, an expanding infestation of spoilage microorganisms can destroy a substantial portion of a stored lot of fruits. P. expansum and B. cinerea are pathogens of apples, pears, and a number of other pectin-rich fruits. B. cinerea is an especially sophisticated and selective plant pathogen that possesses multiple cutinases and lipases that are capable of degrading plants rich in pectin.

The bacterium Erwinia carotovora subsp. carotovora is a highly effective spoilage microbe that causes soft rot across a broad host range of vegetables and some fruits. One of six known genera of soft-rot bacteria (including Xanthomonas, Pseudomonas, Clostridium, Cytophaga, and Bacillus), E. carotovora subsp. carotovora is one of several species of Erwinia that infect and destroy plant tissues both preand postharvest and is the species that causes the greatest damage to harvested vegetables. Soft rot is a form of decay characterized by a watery transparency in infected leafy plant parts and watery disintegration of non-leafy plant materials. "Soft-rot erwinia" tend to initiate infection and decay at wound sites and, once established, can quickly advance to total destruction of the product. Soft-rot erwinia express four pectin-degrading extracellular enzymes: pectin lyase, polygalacturonase, pectin methylesterase, and pectate lyase. Of these enzymes, pectate lyase is primarily responsible for extensive decay. E. carotovora has built-in redundancy for this apparently critical pathogenicity factor, expressing four distinct extracellular pectate lyase isozymes.

Table: Fungal vegetable pathogens[a].

	2004 Annual US per capita consumption (lbs)[b]	Geotrichum	Rhizopus	Phytophthora	Fusarium	Pythium	Alternaria	Colletotrichum	Botrytis	Sclerotinia
Broccoli	5.9		+				+		+	+
Cabbage	8.3		+				+			+
Carrots	8.9	+							+	+
Corn, sweet	9.6				+					
Cucumbers	6.3		+		+	+		+		
Lettuce, head	22.5	+							+	+
Lettuce, leaf	12.0	+							+	+
Mushrooms	2.6									
Onions	21.7	+			+		+	+	+	
Potatoes	46.5		+	+	+	+				
Spinach	2.1									
Tomatoes	19.3	+	+	+	+		+	+	+	+

[a]Important postharvest diseases retrieved from researchers.
[b]81.0% of all fresh vegetables consumed per capita in the United States in 2004.

Soft-rot erwinia are active only at temperatures of 20 °C and above, which reinforces the need to maintain a continuous cold chain from immediately postharvest to retail to successfully manage this ubiquitous spoilage bacterium. Another group of soft-rotting bacteria, the fluorescent pseudomonads (i.e., Pseudomonas fluorescens and Pseudomonas viridiflava), can decay plant tissue at temperatures at or below 4 °C. This is one explanation for the high prevalence of these bacteria on decayed vegetables at wholesale and retail markets. Several P. viridiflava mutants defective in pectate lyase expression that these mutants completely lost the ability to induce soft rot on potato tuber slices. The soft-rotting fluorescent pseudomonads, when considered together with soft-rot erwinia, present a formidable challenge to commercial fresh product operations, and fresh vegetables in particular, from the farm to retail and wholesale outlets.

Pseudomonas tolaasii, another fluorescent pseudomonad and fresh produce spoilage bacterium, has a much narrower range of host-specificity than P. fluorescens and P. viridiflava. P. tolaasii causes spoilage of the white mushroom, Agaricus bisporus. Similar to P. fluorescens and P. viridiflava, P. tolaasii produces siderophores that fluoresce under ultraviolet light. However, unlike the soft-rot pseudomonads, P. tolaasii does not cause soft rot on plants (i.e., it does not produce pectin depolymerases) but instead creates unsightly blemishes on the caps and stems of the Agaricus fruiting body as a result of localized infection and decay of those parts of the mushroom. Wells, Sapers, Fett, Butterfield, Jones, Bouzar, & Miller identified three pathotypes of mushrooms based on pathology and fatty acid analysis: P. tolaasii and P. gingeri which cause severe and yellowed lesions and P. reactans which causes a mild discoloration of the infected area.

Prevention and Control Measures

Pre-Harvest and Harvest Factors

Fresh fruits and vegetables are among the more challenging of food products to commercially produce and distribute. Fresh produce remains metabolically and developmentally active as it proceeds from the commercially appropriate time to harvest (horticultural maturity), to physiological maturity, to senescence and complete deterioration. During this period of development, several physiological and compositional changes occur. This process can be summarized chronologically as growth, maturation, physiological maturity, ripening, and senescence. Although infection and microbiological spoilage can proceed at any time during this developmental continuum, the period of greatest susceptibility to decay onset is during ripening and senescence. Prior to ripening, fruits and vegetables are equipped with defensive barriers to infection including active wound healing and the production of phytoalexins which are phenolic substances that are toxic to fungi.

Losses due to postharvest spoilage or pathological decay are a result either of latent infections in the field that become active following harvest or of crosscontamination during harvest, cleaning, storage, and distribution. Presence of the pathogen on a susceptible host fruit or vegetable, combined with suitable environmental conditions such as high temperature, provides the three components required for disease expression such as host, environment, and pathogen. Therefore, spoilage management should begin in the field using an integrated strategy of GAPs. Balanced crop nutrition influences susceptibility to spoilage. For example, Sugar, Righetti, Sanchez, and Khemira determined that adjusting pear orchard nutrition, specifically for low nitrogen and high calcium, reduced fruit decay postharvest. High nitrogen in plant tissues generally increases susceptibility to decay, whereas high calcium content reduces postharvest decay on several crops. Removing dead and decaying plant matter and other organic material from the crop plant and soil surface will eliminate a major harborage for spoilage microbes as well as other crop pests. To the extent possible, isolating the agricultural field from wild and domestic animals will not only reduce total microbial pressure on the crop, but also reduce food safety risks. Aerial fungicide applications preharvest also will reduce postharvest spoilage in storage. For example, a single application of the fungicide ziram to pome fruit reduced postharvest decay by 25–50%. Other preharvest fungicides are also available (e.g., iprodione and cypronidil) and several new fungicides are under development. In addition, insect pest management will reduce insect damage to crops and also will reduce microbial cross-contamination by the insect vector. This is especially important for chewing insects that create wounds on the fruit or vegetable and can simultaneously inoculate the wound site.

At time of harvest and throughout handling before storage and distribution, it is important to minimize wounds and bruising and to cull all damaged and diseased product. A few spoilage microbes, primarily fungi, can infect healthy tissues by forming appressoria, external structures that enable the pathogen to penetrate the cuticle and epidermis. The developing appressorium ramifies through these protective layers and into the pulp through a combination of mechanical pressure and tissue destruction by extracellular enzymes. However, most spoilage microbes infect and initiate decay at punctures and splits in the epidermal layer or, in far fewer cases, through natural openings such as stomata and lenticels.

Post-Harvest Factors

Product integrity at time of harvest and stringent temperature management from harvest to consumption are two critically important factors contributing to acceptable storage and shelf life of all fresh fruits and vegetables. Upon harvest, fresh fruits and vegetables benefit from immediate surface sanitation and rapid cooling to slow product metabolism and growth of spoilage microbes. Reducing the rate of metabolism likewise reduces product respiration which, in turn, reduces the rate of deterioration, or perishability, of the crop. In many instances, product cooling and sanitation are accomplished simultaneously through one or more washing with chilled water amended with a sanitizing chemical. Chlorine, as sodium hypochlorite, calcium hypochlorite, or chlorine gas, is the most commonly used sanitizing chemical in the produce industry. Chlorine at a rate of between 50 and 200 ppm is added to pre-chilled water which is then applied to harvested fruit as a dip or as a spray or as some combination of these two methods. Concentrations below 50 ppm may not be particularly effective on some fruit, and concentrations above 200 ppm may damage the product and also create a potential worker safety issue due to off-gassing. To achieve and maintain maximum sanitizing efficacy, it is important to maintain water pH at or slightly below neutrality (pH 6.5–7.0). This can be achieved using any of a number of food-grade acids such as citric acid. It is also important to maintain as low an organic load as possible in the wash water because chlorine is unstable in the presence of organic matter and is rapidly inactivated. Other sanitizing chemicals such as ozone, chlorine dioxide, and peroxyacetic acid also are approved for use on fresh produce and are available commercially. Methods for monitoring sanitizer concentration are available for all commercially available sanitizers, and it is also strongly recommended to perform routine treatment efficacy assessments. This can be determined by collecting a minimum of three, and preferably five, individual samples immediately prior to washing and another three or five samples immediately following washing, and determining the total aerobic plate count on each sample. A properly functioning wash system should reduce the average total aerobic plate count by 10to 100-fold. Sequential wash steps will further improve product sanitation by providing greater reductions in microbial load on the product.

Milk and Dairy Products' Spoilage

Consumers currently determine milk spoilage by checking the "sell by" and "best if used by" dates on milk cartons provided by suppliers. These dates are simple estimates of milk shelf life and are often inaccurate due to the variable processing, shipping, and storage conditions of the milk Retailers and consumers discard billions of pounds of unspoiled milk each year while relying on inaccurate printed expiration dates. Conversely, milk may spoil before the printed expiration, and spoiled milk can lead to food poisoning if consumed Another issue demonstrating the urgency of developing more accurate milk packaging is that unless there are local restrictions for dairy products, the manufacturer determines the date; if an established regulation does not exist, manufacturers can legally sell the expired product past the posted date. The cities or processors that do have regulations have different sets of rules. For example, until 2010, milk could only be legally sold in New York City up to 96 hours after 6:00 AM on the day after pasteurization. Dairylea Cooperative, Inc., the biggest processor in the region, allows milk to be sold 10 to 12 days after pasteurization.

Researchers have recently begun investigating applications of current technologies for the detection of milk spoilage. While the meat, fish, and fruit industries have continually advanced new methods of packaging, packaging innovation in the milk packaging industry has remained stagnant. Although a handful of packaging companies have proposed ideas for updated milk packaging designs, none of the proposed alternatives have succeeded in the market. Food industries and consumers have shown a trend of increasing interest in environmentally friendly, health-conscious intelligent packaging, and these market trends indicate the need to invest heavily in these novel developments. Innovative milk packaging will further revolutionize the packaging industry and transform the way consumers think about packaging.

Current Methods of Milk Spoilage Detection

Utilizing pH Indicators as a Measure of Spoilage

Bacteria growth varies from one species of bacteria to another. While one bacteria species may prosper under certain conditions, another species may weaken. These conditions are interdependent and include nutrient availability, moisture, oxygen levels and the level of other gases, the presence of inhibitors, temperature, and pH.

The pH of unspoiled milk is approximately 6.7, a level at which many forms of bacteria thrive. At lower pH levels of 4.0 5.0, lactic acid bacteria can grow and produce lactic acid. While these organisms inhibit the growth of many pathogenic bacteria and are also intentionally employed to ferment milk to make other dairy products such as yogurt and cheese, they can also induce undesirable spoilage in certain products.

Coliforms, a common form of bacteria, have been an indicator of the presence of pathogens in assessing the contamination of water as well as dairy products Coliforms can cause rapid spoilage in milk because they ferment lactose with the production of acid and gas, and they can also degrade milk proteins. Escherichia coli is a well-known example of a coliform Studies have shown that other properties of milk also promote bacteria growth, such as the high availability of moisture and dissolved oxygen which supports both aerobic and facultative anaerobic microorganisms Temperature is frequently controlled to limit bacteria growth. Extreme heat is lethal to many organisms, such as coliforms, which explains the process of milk pasteurization (63 °C for 30 minutes). Two types of bacteria exist in pasteurized milk: thermoduric bacteria, which are capable of surviving the extreme heat during pasteurization, and bacteria that originate from unsanitary conditions postpasteurization Psychrotrophs comprise the largest percentage of bacteria in milk and cause spoilage in refrigerator temperatures at or below 7 °C.

Acidity increases as milk spoils; thus, acidity can be quantified to measure milk quality. Acidity in dairy products can be expressed in two ways: 1) titratable acidity, which shows total acidity but not acid strength; and 2) hydrogen ion concentration or pH, which indicates acid strength. The natural acidity of milk is 0.16% 0.18%, and samples with higher figures indicate developed acidity. At normal levels of pH, the main protein in milk, casein, remains evenly dispersed. At lower levels of pH below 4.6, the protein can no longer remain in solution, so it coagulates due to acid generated from fermentation.

Two studies confirm the link between pH change in milk and spoilage: Fromm and Boor (2004) researched the attributes of pasteurized fluid milk (2% High Temperature/Short Time, HTST milk) during its shelf life. Milk samples were randomly collected from three fluid milk processing plants in the state of New York. A group of 13 panelists evaluated 2% HTST processed fluid milk products based on a quantitative descriptive analysis. They tasted and scored the perceived intensity of the aroma, taste, and aftertaste of milk samples varying in degree of freshness using a numeric scale ranging from 0 to 15 and the descriptive terms listed in Table.

The free fatty acid (FFA) content of the samples significantly increased throughout shelf life. Though not significantly different between day one and day seven, the FFA content drastically increased between days fourteen to seventeen due to milkfat lipolysis. The higher the FFA, the more likely sensory panelists were able to detect lipolyzed or rancid off-flavors in 2% fat milk.

Casein levels in all milk samples also decreased approximately 2% at a relatively rapid rate following seventeen days of refrigeration. This is associated with off flavors in fluid milk, particularly bitterness.

Table: Descriptors of attributes of pasteurized fluid milk by Fromm and Boor.

Descriptors		Reference for intensity of attribute
Category	Attribute	
Aroma	Cheese	Colby cheese
	Cooked	Freshly pasteurized milk (79 °C/18s)
	Cream	Heavy cream
	Hay/grain	Hay
	Sulfur	Boiled egg
	Sour/fermented	Sauerkraut
	Putrid	Limburger cheese
Taste	Baby formula	Baby formula
	Butter	Butter
	Cooked	Freshly pasteurized milk (79 °C/18s)
	Flat	2% fat milk with 3% added water
	Nutty	Peanuts
	Rancid	Strong provolone cheese
	Sweet	1:1 100% lactose free Lactaid and 2% fat milk
Aftertast	Cardboard	Cardboard box
	Sweet	1:1 100% lactose free Lactaid and 2% fat milk
	Sour	Cultured buttermilk
	Metallic	$FeSO4$ (0.445 g/L)
	Drying	Unsweetened tea
	Lingering	-

Each processing plant has different microflora species and needs to have plant-specific strategies to identify and reduce sources of contamination. However, these species, while different, all cause milk to decrease in pH. Increases in FFA and drops in casein levels correlate with a decrease in pH. This suggests that pH can serve as a measurement not only of milk spoilage but also of milk edibility, since panelists determined FFA and casein levels affect rancidity and off flavors in fluid milk This specific study, however, does not establish a lactic acid level and corresponding pH at which milk remains drinkable. Though pH meters are available, modern versions are inconvenient and cumbersome for individual consumers.

Researchers conducted a study to analyze the amount of metabolic products produced by five specific probiotic strains in ultra-high temperature (UHT) treated milk The study used a pH meter with a combined glass electrode and temperature probe to measure pH during fermentation. Volatile compounds were analyzed with headspace gas chromatography and organic acids were analyzed with high pressure liquid chromatography. Quantitative analysis of carbon dioxide production was determined with an infrared CO_2 gas analyzer. Preliminary studies showed that the growth varied considerably with the concentration of the added supplements. After 6-16 hours of incubation, all strains attained viable cell numbers above 8.7-9.18 log CFU/mL. Depending on the probiotic strain used, the pH of the ultra-high temperature milk decreased from 6.7 initially to 3.9-4.4 after 24 hours of incubation.

A disadvantage of the study was that in fortified milk, the various probiotic strains possessed different metabolic profiles, which affected the sensory quality of products containing these different organisms. The increase in strain growth (cell numbers), led to an increase in the amount of lactic acid produced. This increase in lactic acid in turn led to a drop in pH. Whereas the earlier methodology correlated levels of pH with amounts of protein, this study correlated pH directly with the number of bacteria cells.

These studies conclude pH is quantifiable and can measure spoilage in milk. However, there must be further research to address the shift in metabolism of probiotic bacteria in response to environmental changes, as well as the effect of different milk treatment regimes on the metabolism of probiotic bacteria in milk. In addition, the pH range where milk meets the definition of "spoiled" does not coincide with the colloquial definition of spoilage. As in the earlier study, most consumers consider milk to be inedible at those pH levels. For the future development of pH as an indicator of milk quality, a more accurate pH range must be established to define the point at which milk is no longer drinkable.

There are currently a few devices that can determine the acidity levels of milk. These are typically used as a means of quality control by manufacturers, rather than by end consumers. Many pH electrodes/meters and titrators/meters quickly and accurately measure pH of dairy products. A prototype, the Milkmaid smart jug, is a new product that detects when milk starts to spoil with a pH sensor in the base. The jug informs users of spoilage via a change in the color of the jug's light-emitting diode lights. This product is not yet sold and the price has not yet been determined. A disadvantage of this design is that consumers must pour milk into a separate container, because the pH sensor is not incorporated into the plastic container milk is sold in. Thus, while the Milkmaid jug shows that detecting spoilage with pH is applicable to commercial products, it remains to be seen if this particular design becomes commercially successful.

Table: Viable pH counts in UHT milk (log·cfu·ml^{-1})^2.

Time of freezing										
0 hour			1 day			1 month			1 month	
Sucrose (%)										
Strain	0	0	5	10	0	5	10	0	5	10
Lb. acidophilus La5	8.78	8.63	8.70	8.76	8.79	8.84	8.81	8.70	8.79	8.80

Lb. acidophilus 1748	10.11	-	-	-	10.18	-	-	-	-	-
Lb. rhamnosus GG	9.08	9.04	9.06	9.09	9.15	9.05	9.08	9.17	9.16	9.25
Lb. reuteri SD 2112	9.17	9.05	9.08	9.02	9.15	9.05	9.08	9.17	9.16	9.25
B. animalis BB12	8.86	8.84	8.83	8.80	8.93	8.93	8.87	8.75	8.71	8.76

Electrical Methods for the Detection of Bacteria

Some traditional methods of detection involve bacterial enumeration, in which spoilage is detected when increased metabolism caused by multiplying bacteria renders a colored solution colorless. The methylene blue reduction test is such an example; however, known flaws of this test include time-consuming and redundant procedures, as well as an inability to discriminate between bacterial types. Researcher sought to improve upon the methylene blue reduction method while maintaining its advantages by supplementing it with an amperometric sensor. An amperometric sensor, composed of a circuit with a potentiostat and a pair of electrodes, measures current change Amperometric sensors are small and inexpensive and have been tested in a variety of media to detect changes in bacteria such as E. coli.

E. coli and Ent. aerogenes, two types of coliforms that indicate the sanitary condition. A third sample contained milk and methylene blue. Methylene blue is blue until the metabolic activity of bacteria causes it to lose color. Consequently, the bacterial metabolism of the E. coli caused the reduction of methylene blue in the three samples and also resulted in a current change. Any current change of more than 0.05 μA was detected with the amperometric sensor and recorded. The study tracked detection time and provided an estimate of the approximate number of microorganisms initially in the sample. Results had R^2 of 0.9192, corroborating high accuracy in an inverse linear relationship between the log of the bacterial concentration against the detection time. The increase of microbial organisms exponentially related to the time from inoculation to the initial small change in current.

Results were favorable. Advantages to this method include a detection time 0.5-2 hours shorter than that obtained with the methylene blue reduction method and a very broad detection range of 10^2-10^4 CFU/mL. Furthermore, whereas the methylene blue reduction method required constant supervision and sampling at a 30-minute interval, the amperometric sensor could independently record the data. The latter procedure was relatively simple and inexpensive; accuracy was also a non-issue.

However, this method cannot discriminate between viable and non-viable cells. Furthermore, type of bacteria detection was lacking. The amperometric sensor could only detect E. coli and Ent. aerogenes coliforms; when other bacteria such as B. subtilis, Lactobacillus sp., Saccharomyces sp., and Staph. aureus were tested upon, they produced a negligible current change.

Wireless Detection and Monitoring of Milk Spoilage

Application of remote-query technology to detect milk spoilage is an emerging field of experimentation. The remote-query magneto-elastic sensor platform is a free standing, ribbon-like magneto-elastic thick-film coupled with a chemical or biochemical sensing layer such as an enzyme that vibrates at a characteristic resonance frequency. A pickup coil is then used to remotely detect the magnetic field generated by the mechanical oscillations. Magneto-elastic sensors have already been developed and tried in a number of different types of analyses; past research has used it for the analysis of glucose concentration, blood clotting, and detection of Escherichia coli as well as Salmonella typhimurium However, it has not directly been applied to the detection of milk spoilage.

Researchers tested a remote-query magnetoelastic sensing platform to quantify the bacterial count of Staphylococcus aureus ssp. anaerobius (S. aureus) in milk. S. aureus is a bacterium that resides in milk and multiplies as milk spoils; infection with S. aureus can result in such human diseases as toxic shock syndrome, endocarditis, and septicemia. Culture media of S. aureus were prepared, and resonance characteristics were measured with a magnetoelastic sensor fabricated from Metglas alloy ribbon. Because the sensor responds to mass loading due to bacterial adhesion and changes in solution viscosity, researchers increased the viscosity of one of the culture media—one a nutrient broth and one a tryptic soy broth—through different trials; ultimately, the sensor showed a higher sensitivity in the milk than in the culture medium because of the higher viscosity of milk. Results concluded that the sensor platform is feasible for use in the remote detection of spoiled milk samples.

Remote-query detection is, overall, an effective method. Typically, material costs are low; disposable magnetoelastic sensors, for example, can be fabricated from strips costing $300 per kilometer, while the combination of an amperometric sensor with methylene blue also involves low costs. However, deficiencies in this method continue to surface. As bacteria detection becomes complex, as with the ATP (adenosine triphosphate) bioluminescence method and the PCR (polymerase chain reaction) method, speed becomes an issue. The simple methylene blue reduction method was ponderous as bacteria detection requires constant supervision.

The two methods presented attempts to improve upon these inefficiencies. Not only does researchers experimentation with the amperometric sensor provide a broad detection range between 10^2-10^4 CFU/mL, electrochemical techniques are easier to apply and have lower costs.

Gas-Sensor Arrays

Researchers attempted to predict the shelf-life of milk with gas-sensor arrays. In their experiment, a commercial solid-state gas-sensor array system monitored the growth of disinfectant-resistant bacteria in milk known to cause spoilage, namely Serratia marcescens, Serratia proteamacufans, and Pseudomonas putida.

The researchers study identified the quality and amount of major volatile microbial metabolites using gas chromatography and mass spectroscopy, and also analysed and correlated microbiological data through statistical tests. Patterned and fingerprinted complicated odors with gas sensor arrays in order to determine the correlation between secondary volatile metabolites from the milk and the microorganisms that produced these compounds. This method allows the determination of organisms responsible for spoilage, quality of the milk, and length of shelf life based on objective criteria Twenty-six Gram-negative bacteria were selected and identified based on cellular fatty acid analysis, and three spoilage-inducing isolates (Cedcea sp., Serratia sp., and Pseudomonas sp.) that proliferate in milk were chosen for experimentation.

Aside from the CO_2 signal, which stayed constant at baseline level, the sensor signals decreased during the first 6-7 hours because background volatile compounds from the medium were extracted dynamically from the cultures during each measurement. The production of volatile secondary bacterial metabolites started to exceed the background volatiles and increased significantly around 7-8 hours (excluding the pure Pseudomonas culture). This increase in signal coincided with the onset of the exponential growth phase. At 18-20 hours, a peak in CO_2 production corresponded to a plateau in S. marcescens N9, S. proteamaculans O6A and the mixed cultures. The sensor signals from the pure cultures showed significant correlation with the cell counts.

This gas-sensor array system successfully detected major volatile metabolites produced by bacteria during growth. A significant majority of the contributions to the sensor response signals were ascribed to the major volatile compounds except for acetate. Researchers investigated the possibility of developing a system that detects and monitors growth of undesirable bacteria. The high correlation between the sensor readings and the cell counts of the pure cultures suggests that gas-sensor measurements can predict bacterial cell numbers in pure cultures. Therefore, a design of specific sensors can be adapted to follow the development of spoilage organisms in specific food products.

Secondary compounds identified by gas chromatography and mass spectroscopy comprised of fermentation products produced by Serratia, Enterobacter, and Erwinia. Both Enterobacteriaceae and Ps. putida produced spoilage off-odors in pasteurized milk, but the Ps. Putida culture contributed a far lower production of volatiles and CO_2 than the Serratia strains at 25 °C. At less than 10 °C, Ps. putida would have been the major spoilage organism in milk. Therefore, sensors must be selected and developed to provide indicators of typical target organisms in different food products under specified temperature, pH, and water activity conditions.

However, there are constraints to detecting bacterial strains in complex cultures based on temporal gas sensor response patterns from single strains. In particular, Ps. Putida produced secondary volatile metabolites and only contributed to less than 0.1% of the temporal response pattern, even though the Ps. putida reached a significant concentration in the mixed culture. Therefore, for the applied experimental conditions, the gas sensors were not sufficiently sensitive to detect Ps. putida in the mixed culture, because the vapor phase was masked by the volatile metabolites generated by S. marcescens N9. The sensors also have a certain extent of cross-sensitivity, so it is important to select sensors with high sensitivity for detection of strain specific metabolites, especially when the spoilage bacteria constitute a lesser percentage of the total flora.

Gas sensors can be designed to detect the characteristic metabolites of specific bacteria in specific food products and can offer rapid, accurate determinations of shelf life. To detect correctly target strains of bacteria, it is necessary for sufficient strain specific volatile compounds to be present. Thus, a drawback of gas sensor technology is its great potential for strains where the strain specific metabolites represent the major volatile compounds in the vapor phase. When bacteria do not produce many volatile compounds, more sensitive sensors are needed to detect their characteristic metabolites.

Infrared Spectroscopy as Spoilage Indicator

Spectroscopy is a non-destructive technique, where spectral features provide biochemical information regarding the molecular interactions between, and the composition and structure of, different cells and tissues. This method was first widely applied in the food industry to detect spoilage in beef, rainbow trout fillets, and other meat products. However, this method had not been experimented on milk until Al-Qadiri, M. Lin, Researchers in 2008. To do so, they evaluated visible and short wavelength near-infrared diffuse reflectance spectroscopy (SW-NIR) as a technique to detect milk spoilage in pasteurized skim milk. They wanted to see the feasibility of applying visible and SW-NIR spectroscopy to monitor spoilage of pasteurized skim milk in industrial settings.

In doing so, Researchers first took the total aerobic plate count and pH measurements. They then examined the milk samples at 22 °C to correct for spectral changes that could result from temperature differences during spectral collection. The mean pH measurement for control milk samples was 6.66, and they found no obvious pH decrease for milk samples stored at 6 °C after 30 hours of storage. In experimental samples, the visible and SW-NIR diffuse spectroscopy detected the formation of metabolic by-products from proteolysis and lipolysis caused by bacterial cell growth, which led to a reduction in pH. This method was effective, but costly. Further work will be needed to identify which biochemical changes in spoilage micro-organisms correlate with specific SW-NIR spectral features.

Researchers attempted to take infrared spectroscopy further with matrix-assisted laser desorption/ ionization time-of-flight mass spectrometry (MALDITOF-MS). MALDI-TOF-MS has already been used in protein and peptide identification and quantification; however, researchers wanted to see if it was useful for microbial spoilage assessment because techniques for identifying and quantifying spoilage bacteria in pasteurized milk are time-consuming. Their methodology included incubating milk samples and raw pork meat samples at 15 °C and at room temperature, and then analysing them with MALDI-TOF-MS at a rate of 4-minute intervals.

MALDI-TOF-MS has many advantages, particularly in terms of sensitivity, accuracy, and speed. Spectrum can be generated within minutes following sample preparation. It's most comparable technology is fourier transform infrared (FT-IR) spectroscopy; however, MS allows more equivocal identification of important proteins while FT-IR spectroscopy does not, or does so at best only empirically through peak assignments Furthermore, MALDI-TOF-MS has minimal sample preparation, which contributes to the rapid speed of data collection. Typical sample speed is 4 minutes per sample, which is considerably faster than classical microbiological plating approaches that can take up to 2 days. Drawbacks, however, include the limited use of infrared spectroscopy in the field. The technology is perceived as a tool for assessing protein qualitatively rather than for measuring microbial bacterial count quantitatively. Familiarization of MALDI-TOF-MS can help change perceptions and lead to use of this technology in the dairy industry. However, the technical difficulty of this method renders it unsuitable for consumer use.

Protein or Fat Count Detection

Several studies correlate bacteria that are known to be involved in the spoilage of pasteurized milk with variations in the levels of lipids and proteins that are present in milk. The lowest percentages of lipid and protein in milk occur when Pseudomonas levels are at their highest. High Pseudomonas levels cause high levels of proteolytic activity in all food systems. Lipases and proteases break down lipids and proteins, which lowers lipid and protein levels. The by-products of these hydrolysis reactions increase milk acidity and directly correlate to milk spoilage. Additionally, the bacteria Pseudomonas aeruginosa change the constituents of milk, thereby spoiling it.

There are many common industry practices that incorporate this method into testing milk properties. Protein levels are typically determined by the standard Kjeldahl method or the more favorable

Dumas method The Kjeldahl method measures nitrogen levels by using a fitting titration technique. There are three steps involved: digestion, neutralization, and titration. The protein content is calculated from the nitrogen concentration in the milk. This standard method does not directly measure protein content and thus needs a conversion factor (which varies among different proteins) to convert measured nitrogen concentration to a protein concentration. This conversion can lead to inaccuracies. Another disadvantage is the amount of time (1-2 hours) required to perform this test.

The Dumas method, a similar but enhanced version of Kjeldahl method, is an automated instrumental technique that combusts in the presence of oxygen a sample of known mass in a high temperature chamber. By-products CO_2 and H_2O are filtered out, leaving only N_2 or nitrogen content to be read by a thermal conductivity detector. This test is can be done in fewer than 4 minutes. Like the Kjeldahl method, the Dumas method also needs to convert nitrogen content to protein content, as it does not truly measure the protein. The method also carries high initial costs and can make it difficult to obtain a representative sample. Alternate methods of protein counts are time consuming and require extensive sample preparation before analysis.

Lipid or fat content is primarily determined by Gerber method in Europe, or the very similar Babcock method in the United States. The Gerber method requires adding dairy product into a butyrometer and adding concentrated sulfuric acid and amyl alcohol to dissolve the non-fat milk solids. The mixture is centrifuged for a set time at 1100 rpm and placed in a water bath to standardize the samples before reading the fat content off the calibrated scale of the butyrometer. The method is fast, low cost, and suitable for high volumes of sample. However, disadvantages include not being able to automatically determine levels and the risk of handling highly concentrated sulfuric acid. In addition, reading the butyrometer takes acquired skill, so the method cannot be used by average consumers.

Sorhaug and Stepaniak showed that psychotropic Bacillus spp. are resistant to heat treatment and secretes extracellular proteinases, lipases, and phosoplipases. Other bacteria that survive the pasteurization process also produce these proteins. Additionally, the quality of dairy products can decrease due to heat-resistant enzymes that are secreted by psychrotrophs in raw milk before heat treatment, or produced by psychrotrophs growing during the cold refrigeration of dairy products Prior studies also show that up to 20% of all psychrotrophic bacteria isolated from raw milk possess proteolytic and lipolytic enzymatic activities Sorhaug and Stepaniak determined the average psychrotroph count for different dairy types at the time of spoilage for Pasteurized milk to be 6-7.5 CFU/mL, but they did not record the protein level. For this method to be used as a spoilage detector, more testing is needed to determine the equivalent protein level at the psychrotroph count of 6-7.5 CFU/mL and the precise correlation between protein count and spoilage.

Average consumers cannot perform a protein or fat count due to the method's required know-how, needed time, and financial costs. Thus, it is impractical to incorporate this method into commercially sold products for consumers, even in light of the fact that levels of post-pasteurization contamination correlate extremely well with shelf life, and protein and fat count is arguably the most accurate indicator of expected shelf life of pasteurized milks for milk processors. Researchers, also determined that the by-products of lipid and protein breakdown induce increased acidity.

Comparison of Milk Spoilage Detection Methods

While the milk packaging industry has lagged woefully behind in milk spoilage detection, there has been a recent push towards innovation in regards to spoilage prevention. These efforts do not eliminate the need for future research into milk spoilage detection methods, but could serve as substitute solutions to the problems previously discussed in this paper. Detailed below are some examples of these recent innovations.

Tetra Pak USA

Tetra Pak is a leading food processing and packaging solutions company in the world that emphasizes innovation and environmental friendliness in their products. In the general food industry, they have introduced the Tetra Recart, which is the first retortable carton package designed for shelf-stable products traditionally filled in cans, glass jars, or pouches. The Tetra Recart guarantees freshness for up to 24 months. This product is currently available for use only with shelf-stable products, but the company has announced further plans to begin development on a similar type of carton for milk which would guarantee freshness for up to 6 months.

Evergreen Packaging, Inc.

Evergreen Packaging, Inc. specializes in dairy, juice, and liquid packaging and caters to customers specific needs. Evergreen Packaging focused their milk carton design on preventing light and temperature changes since dairy products are sensitive to these changes and milk reacts negatively to increases in temperature and light exposure. Evergreen Packaging claims their fiber-based packaging keeps light and oxygen out while retaining vitamins and taste, which helps lock in freshness. Specifically, they have patented their superior oxygen and moisture-barrier technology.

Chapter 3

Fermentation and Fermented Foods

Fermentation refers to the metabolic process of breakdown of an organic substrate by the microorganisms through the action of enzymes. Fermented foods can be divided into animal-based fermented foods, plant-based fermented foods and fermented drinks. The topics elaborated in this chapter will help in gaining a better perspective about fermentation and fermented foods.

Fermentation

The most common meaning of fermentation is the conversion of a sugar into an organic acid or an alcohol. Fermentation occurs naturally in many foods and humans have intentionally used it since ancient times to improve both the preservation and organoleptic properties of food. However, the term "fermentation" is also used in a broader sense for the intentional use of microorganisms such as bacteria, yeast, and fungi to make products useful to humans (biomass, enzymes, primary and secondary metabolites, recombinant products, and products of biotransformation) on an industrial scale.

Modern industrial fermentation processes used in the food and beverage industry can be described according to different perspectives. In the centre of these processes are usually bioreactors, which can be classified with respect to the feeding of the bioreactor (batch, fed-batch, and continuous mode of operation), immobilization of the biocatalyst (free or immobilized cells/enzymes), the characteristic state of matter in the system (submerged or solid substrate fermentations), single strain/mixed culture processes, mixing of the bioreactor (mechanical, pneumatic, and hydraulic agitation), or the availability of oxygen (aerobic, micro-aerobic, and anaerobic processes). The decision as to which bioreactor or fermentation process should be implemented in any particular application involves considering the advantages and disadvantages of each setup. This includes examining the properties and availability of the primary raw materials, any necessary investment and operating costs, sustainability, availability of a competent workforce, as well as the desired productivity and return on investment. Since in large-scale applications, each fermentation system needs to operate efficiently and reliably, the major criterion for the selection of a bioreactor/fermentation process remains the minimum for capital costs per unit of product recovered. Simultaneously, with efficient design and operation, in large-scale processes, the issues concerning by-product and wastewater management are inevitable.

Types of Fermentation Processes

Submerged Cultivation

Submerged cultivation of microbial cells in bioreactors guarantees a controlled environment for the efficient production of high-quality end products and to achieve optimum productivity and yield. Industrial bioreactors operated in batch, fed-batch, or continuous mode are utilized to culture different types of microorganisms producing a wide range of products. In the following sections, different approaches to submerged cultivation of microorganisms in bioreactors are discussed briefly and the typical features, benefits, and drawbacks of each cultivation mode are highlighted. Finally, the relevant applications for batch, fed-batch, and continuous cultivation of microorganisms in liquid media used in the production of different types of food industry products are demonstrated.

Batch Cultivation

Batch culture represents a closed system in which the medium, nutrients, and inoculum are added to the bioreactor, mostly under aseptic conditions, at the beginning of cultivation that is, the volume of the culture broth in the bioreactor is theoretically constant during cultivation (practically, small deviations in culture volume are caused by a low feed rate of acid/base solutions to keep the pH at a desired level and by sampling or introducing air/gas into the culture; on balance, such changes are usually ignored due to their small value relative to the total working volume of the bioreactor).

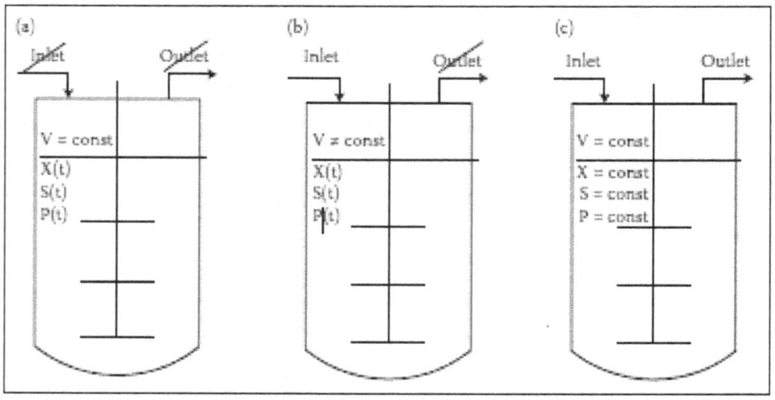

Simplified scheme of (a) batch, (b) fed-batch, and (c) continuous cultivation.

Typically, at the beginning of batch cultivation, a known number of viable cells are inoculated into the bioreactor that is already filled with sterilized medium containing all nutrients. After inoculation, the cell culture follows the classical growth curve described by Monod, which is divided into four main phases. As the lag phase is an "inefficient" stage of culture (even though the cells are metabolically active—they are adapting their enzymatic apparatus to a new environment; no significant increases in biomass concentration, substrate consumption, or product synthesis are observed), it is desirable to shorten it as much as possible. The length of the lag phase is influenced mainly by the concentration of cells in the inoculum and their physiological state, the composition of the inoculation and cultivation medium (mainly the source of carbon and energy, pH, and temperature), and the size of the inoculum. The exponential (logarithmic growth) phase is characterized by rapid cell proliferation (biomass concentration is an exponential function of time), constant specific growth rate, which is equal to the maximum specific growth rate of the culture under conditions of absence of growth limitation (growth rate is not limited because all nutrients are present in excess, while also not attaining growth-inhibiting concentrations), fast consumption of the source of carbon and energy, and a high rate of primary metabolite production.

The depletion of nutrients by the end of the exponential phase (in the case of aerobically grown cultures, these are signaled by a rapid increase in dissolved oxygen concentration) causes a progressive reduction in the specific growth rate and a transition to the stationary phase, characterized by the stagnation of growth and utilization of endogenous reserves of carbon and energy; this phase is important for the synthesis of secondary metabolites. Most industrial bioreactors are operated in batch mode due to the relative simplicity of this process. The whole batch operation consists of several steps, including medium formulation, filling the bioreactor, sterilization in place (SIP systems), inoculation, cultivation, product harvesting, and bioreactor cleaning in place (CIP systems). For efficient performance of batch operation, it is important to minimize all non-productive steps (all steps listed above except cultivation), achieve a high rate of product synthesis, optimize productivity, and maximize the yield of the end product. The performance of any particular batch operation is thus influenced by the type of end product—an extension of exponential growth is advantageous for the efficient production of biomass (baker's yeasts, feed biomass) or primary metabolites (ethanol, acetic, citric, or lactic acids), whereas in the case of secondary metabolite production, the exponential phase is shortened (by the limitation of one nutrient, usually the source of nitrogen) and the stationary phase is prolonged to achieve the maximum yield of the product.

Submerged batch cultivation can be used for the production of alcoholic beverages (beer, wine, and distilled spirits such as whisky, brandy, rum, and others), organic acids used in the food industry either as acidifiers or as preservatives (citric, acetic (vinegar), and lactic acids), and amino acids used as flavor enhancers (e.g., monosodium glutamate) or sweeteners (e.g., aspartate).

For distilled spirits, the fermentation of wort during Scotch whisky production is taken as an example. Washbacks, simple cylindrical fermentation vessels (volume 250–500 m^3) for the production of distilled spirits are made either from wood or from stainless steel. Although wood washbacks are difficult to clean and sanitize, they are still used, especially in malt whisky distilleries. Wort to be fermented is pumped to the washback, cooled to 20 °C, and inoculated with either fresh or dried yeast cells.

The global production of citric acid reached 1.8×10^6 tonnes in 2010 (F.O. Licht data); 90% of this was produced by microbial (Aspergillus niger) synthesis from sugar- or starch-containing materials (sugar beet, sugarcane molasses, and corn) and about 60% of this amount was consumed in the food industry. Although citric acid can be produced at an industrial scale using surface liquid cultivation, solidstate cultivation, or submerged liquid cultivation, nowadays, the latter predominates. Submerged cultivation is carried out in stirred bioreactors (capacity 150–200 m^3) or bubble columns (capacity up to 1000 m^3), usually operating aerobically for 4–10 days until the citric acid concentration reaches 10–15% w/v.

Fed-Batch Cultivation

Fed-batch culture represents a semi-open system in which one or more nutrients are aseptically and gradually added to the bioreactor while the product is retained inside; that is, the volume of the culture broth in the bioreactor increases within this time. The main advantages of fed-batch over batch cultures are: (a) the possibility to prolong product synthesis, (b) the ability to achieve higher cell densities and thus increase the amount of the product, which is usually proportional to the concentration of the biomass, (c) the capacity to enhance yield or productivity by controlled sequential addition of nutrients, and (d) the feature of prolonged productive cultivation over the "unprofitable periods" when the bioreactor would normally be prepared for a new batch.

Fed-batch is advantageously used in processes (a) where substrate inhibition or catabolic repression is expected; this problem can be overcome by using a "safe" concentration of the substrate in batch mode followed by feeding the remaining substrate within fed-batch operation, (b) where a Crabtree effect (repression of yeast respiratory enzymes by high concentrations of glucose) is expected by gradual feeding of the substrate, the production of ethanol by yeasts can be eliminated under aerobic conditions, (c) where a high cell density is required; a high and constant specific growth rate can be maintained by exponential feeding of the substrate, (d) where a high production rate should be achieved; cell metabolism can be regulated by precise sequential feeding of nutrients, and (e) where a high viscosity of culture broth is expected (e.g., production of dextran or xanthan); a gradual dilution of the medium can overcome the problems of mixing and oxygen transfer.

There are many methods of adding a substrate to the bioreactor (either as a concentrated solution of a sole carbon and energy source or as a medium containing carbon plus other nutrients); the proper choice of the nutrient feeding rate can enhance the culture performance considerably since it influences cellular growth rate, cell physiology, and the rate of product formation. The common feeding strategies are: (a) discontinuous feeding, achieved by regular or irregular pulses of substrates and (b) regular continuous feeding of nutrients designed according to a pre-calculated profile or based on the feedback control of online measured variables associated with cell growth and metabolism, for example, dissolved oxygen concentration, pH, CO_2, evolution rate, and biomass concentration.

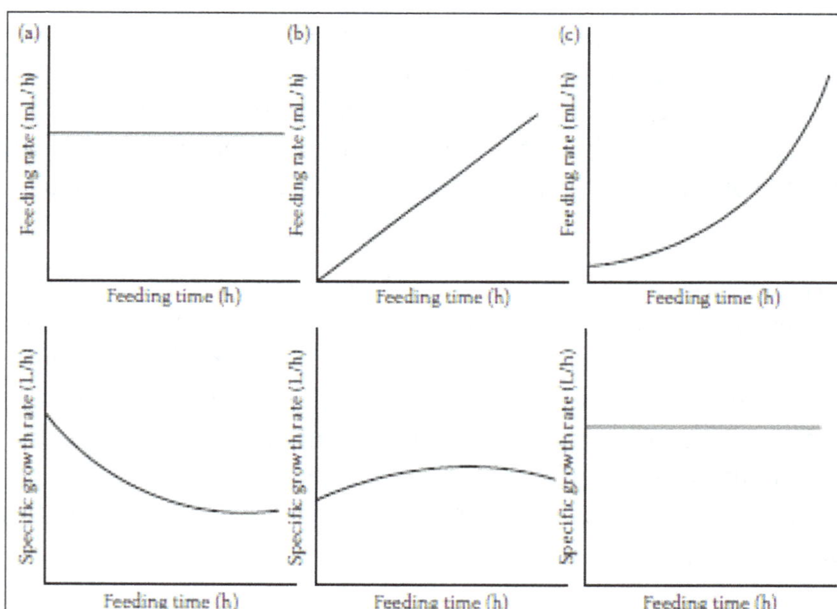

Relationship between pre-calculated profiles of substrate feeding and specific growth rates of cell culture. (a) Constant feeding rate, (b) linearly increasing feeding rate, and (c) exponential feeding rate.

The typical food fed-batch fermentations are large-scale production of baker's yeast, pure ethanol, which is further utilized for alcoholic beverages produced by mixing ingredients such as liquors or cordials, and submerged acetification for vinegar production. Baker's yeast

(Saccharomyces cerevisiae), which is distributed as compressed, dried, or instant biomass, is valued for its dough-leavening ability.

Yeast production is the only technology in which the respiratory metabolism of S. cerevisiae, leading to high biomass yield, is stressed. Nevertheless, due to the Crabtree effect, that is, the formation of ethanol under aerobic conditions in the presence of excess substrate, the only alternative for producing baker's yeast is fed-batch cultivation. The main bottleneck of large-scale baker's yeast production is the control of nutrient medium inflow, which was traditionally based on empirical data. Currently, many feeding methods utilizing different approaches have been developed, for example, a logistic feeding profile evolutionary optimization of genetic feeding algorithms, fuzzy control, pulsed feeding optimization and others.

In the European Union (EU), the annual production of vinegar, comprising 10% (w/v) acetic acid, is estimated to be about 5×10^6 hL. Human knowledge of vinegar production dates back to ancient history and several distinct production methods have been used, for example, surface oxidation (Orleans process, slow vinegar production), the quick vinegar process using a trickle-bed bioreactor, or submerged acetification. The most modern process is submerged acetification, in which acetic acid (vinegar) is produced by Acetobacter-mediated oxidation of ethanol, and takes place in special "acetator" bioreactors. The most common acetator, a Frings Acetator, differs from the usual bioreactors by using a special rotor/stator turbine aerator. The aerator with a double function (aeration and mixing), consists of a rotor placed under the bioreactor and connected to an air-suction pipe surrounded by a stator. The aerator is self-respiring, that is, during rotation (speed about 1500 rpm), it sucks air and pumps liquid, which causes the formation of an air–liquid mixture that is radially injected into the culture medium. The foam is broken by a mechanical defoamer and because oxidation is an exothermal process, cooling is necessary. The acetators are operated in repeat fed-batch mode and one cultivation cycle in a single acetator can produce vinegar containing 15% (w/v) of acetic acid. In addition, a dual-stage high-strength fermentation process, which allows the culture to generate up to 20.5% (w/v) acetic acid, has been developed.

Continuous Cultivation

Continuous culture represents an open system, in which nutrients are aseptically and continuously added to the bioreactor, and the culture broth (containing cells and metabolites) is removed at the same time, that is, the volume of the culture broth is constant due to a constant feed-in and feed-out rate.

Frequently, continuous culture is used as a synonym for a chemostat, represented by a constant specific growth rate of cells, which is equal to the dilution rate and is controlled by the availability of the limiting nutrient, although other types of continuous operation such as turbidistat (a constant concentration of biomass controlled by the dilution rate) or nutristat (a constant parameter related to cell growth controlled by the dilution rate) can be employed. The main advantages of continuous culture (chemostat) over the batch mode are (a) the possibility to set up optimum conditions for maximum and long-term product synthesis, (b) the ability to achieve stable product quality (the steady state is characterized by a homogeneous cell culture represented by a constant concentration of biomass and metabolites), and (c) a distinct reduction in "unprofitable" periods of the bioreactor operation.

In spite of these advantages, there are also several problems that hamper the extensive utilization of continuous operation on a large scale. These include (a) increased risk of contamination due to the pumping of the medium in and out of the bioreactor, (b) the danger of genetic mutations in the production strain in a long-term operation, and (c) additional investments may be required for technical facilities.

Despite the producer's proclamations, only a few food or feed production systems employing microorganisms are operated in a genuine continuous mode, where continuous fermentation is defined as a process running in one or more bioreactors at a stable dilution rate. One of the rare examples of this is fodder yeast (Kluyveromyces fragilis) production using spent sulfite liquor as the substrate, which is operated in the Czech Republic. The main reason why this process can be performed continuously is the low sensitivity of the production medium to possible contamination. At the same time, this process can be taken as a rare example of industrial-scale production using a highly valued substrate, lignocellulosic hydrolysate, which is obtained as a waste product from pulp production.

In other food applications such as in modern distilleries, a semi-continuous fermentation operated in a series of fermentors is usually used instead of genuine continuous fermentation. The current estimates are that only about 16% of ethanol in North America is produced in a continuous mode due to problems with contamination. Nevertheless, both batch and semi-continuous mode fermentations permit continuous distillation, which is the reason why the whole distillery production is often considered as continuous.

Continuous fermentation systems based on immobilized cell technology have also been studied in beer production. Although continuous beer fermentation has been tested as a promising technology for several decades, the number of industrial applications is still limited. The reasons include engineering problems (excess biomass and problems with CO_2 removal, optimization of operating conditions, clogging, and channeling of the reactor) and unrealized cost benefits (carrier price, complex, and unstable operation). The major obstacle hindering the extensive industrial exploitation of this technology is the difficulty in achieving the correct balance of sensory compounds in a short time typical for continuous systems. However, recent developments in reactor design and in our understanding of immobilized cell physiology, together with applications of novel carrier materials, could provide a new stimulus to research and potential applications of this promising technology.

Solid Substrate Fermentation

The term "solid substrate fermentation" (SSF) or "solid substrate cultivation" (SSC) is used for systems where microorganisms are cultured on the surface of a concentrated water-insoluble substrate (usually containing polysaccharides as a carbon and energy source) with a low level of free water. This technique was developed in the Eastern countries, where it has been used for centuries for the production of traditional foods such as soy sauce, koji, miso, or sake, using different substrates and microorganisms. In the Western countries, it has not been widely exploited and its application is limited mainly to the production of industrial enzymes, certain food products, or feed supplements. In the following section, typical features, advantages, and problems of SSF together with relevant applications in both the food and feed industries are discussed.

Typical Features of Solid Substrate Fermentation

Solid substrate fermentation is characterized by very low water activity (the relative humidity of the gaseous phase in equilibrium with the moist solid is significantly below; thus, the main features of this system are substantially different compared with classical submerged cultivation.

There are several advantages of SSF over the conventional submerged technology such as (a) the use of a concentrated medium, resulting in a smaller reactor volume and lower capital investment costs, (b) the lower risk of contamination with yeasts and bacteria due to low moisture levels and substrate complexity, (c) the simplicity of the technology and low production of effluent water from the process, (d) the higher product yield and easier product recovery, and (e) the use of agricultural wastes as substrates for certain applications (e.g., feed supplements and cellulolytic enzymes).

Microorganisms and Substrates Used in SSF Processes

Filamentous fungi are preferable for SSF processes, mainly due to their abilities to (a) grow on substrates with reduced water activity, (b) penetrate their hyphae into the solid substrate, and (c) produce exoenzymes (e.g., amylolytic and cellulolytic enzymes), which decompose the polysaccharides (the main carbon source often present in solid substrates). When using yeast, it is necessary to integrate a material pre-treatment step (such as steam explosion, acid or alkali treatment, followed by enzymatic digestion or a combination of these) into the process or to use a mixed culture, wherein the complex cellulosic or starchy material is first degraded by other organisms (usually molds) that are able to produce extracellular enzymes and the released glucose is then consumed by yeasts (or less frequently by bacteria) yielding the desired product.

Table: Main Differences between Solid Substrate Fermentation and Submerged Cultivation.

Solid Substrate Fermentation	Submerged Cultivation
Low water content of the cultivation medium (40–80%)	Liquid cultivation medium (~95% water content).
Three-phase system: gas–liquid–solid.	Two-phase system: gas–liquid.
Complex substrate insoluble in water, high local concentration of nutrient.	Nutrients are dissolved in water, concentration of nutrients is lower.
Nonhomogeneous system, gradient of nutrients.	Homogeneous system.
Microorganisms are grown on the surface of the solid substrate.	Microorganisms are grown in the liquid medium.
Gas–liquid and liquid–solid oxygen transfer.	Gas–liquid oxygen transfer.
Limitations in heat, oxygen, and nutrient transfer.	Transport processes are usually not limited (exception can be oxygen transfer).
Heat is removed by using a stream of air or by placing the bioreactor into the temperature-controlled chamber.	Cooling is achieved by bioreactor jacket cooling system.
Process monitoring and control are difficult.	Online monitoring and control of the process are common.
High concentration of the product.	Product is dissolved in the liquid phase.

The efficiency of SSF is highly influenced by the selection of the solid substrate. The substrates suitable for SSF should ideally meet the following requirements: (a) have a porous solid matrix with a large surface area per unit volume (10^3–10^6 m^2/cm^3), (b) should sustain gentle compression and mixing, (c) should contain biodegradable carbohydrates, (d) its matrix should absorb water in the proportion of 1 to several times its dry weight, (e) should have relatively high water activity on the solid/gas interface to support microbial growth, and (f) should absorb the additionally added nutrients such as nitrogen sources (ammonia, urea, and peptides) and mineral salts. On the basis of these characteristics, the ideal material consists of small granular or fibrous particles that do not break or stick together. The commonly used solid substrates include wheat, wheat bran, soybean, rice barley, oats, and other cereals.

Animal-Based Fermented Foods

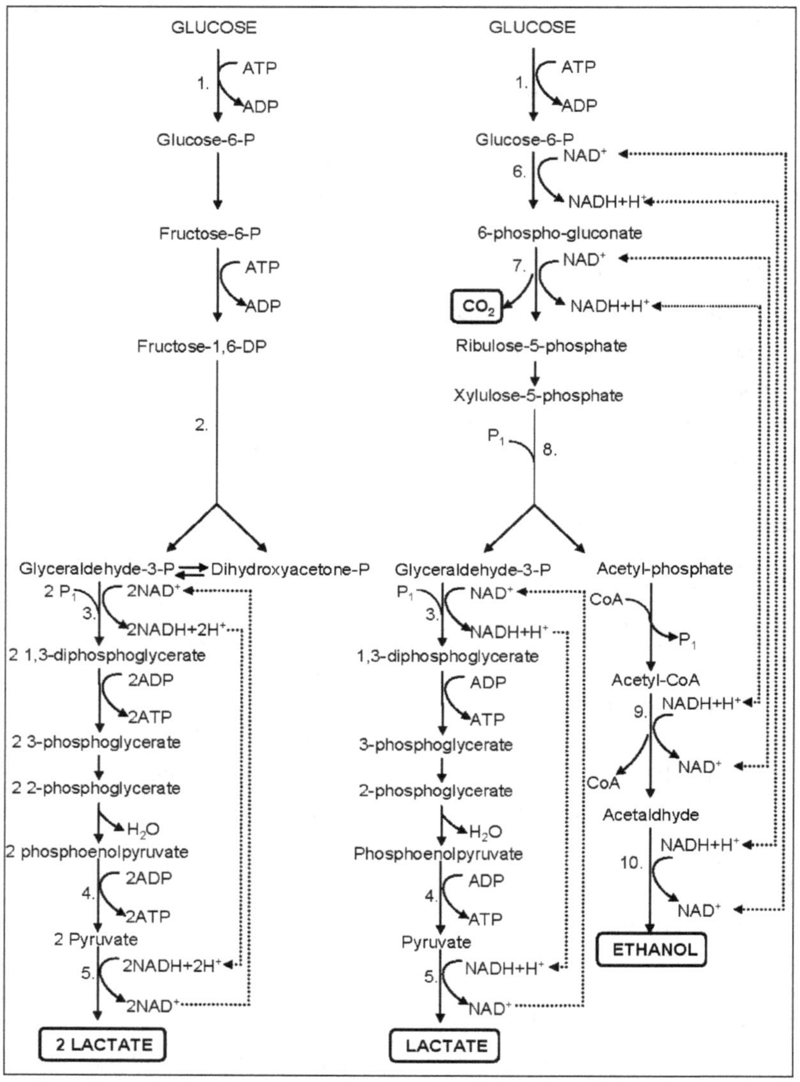

In figure major fermentation pathways of glucose: (A) homolactic fermentation (glycolysis, Embden-Meyerhof-Parnas pathway); (B) heterolactic fermentation (6- phosphogluconate/phosphoketolase pathway). Selected enzymes are numbered: 1. Glucokinase; 2. fructose-1,6-diphosphate aldolase; 3. glyceradehyde-3-phosphate dehydrogenase; 4. pyruvate kinase; 5. lactate dehydrogenase; 6. glucose-6-phosphate dehydrogenase; 7. 6-phosphogluconate dehydrogenase; 8. phosphoketolase; 9. acetaldehyde dehydrogenase; 10. alcohol dehydrogenase.

Fermentation is a simple and inexpensive method for preservation of meat and meat products since antiquity. It has an added advantage of creating of specific products with good aroma. Acid production (lowering the pH), H_2O_2 production and bacteriocins produced either singly or in combination by starter cultures are responsible for preventing the growth of food-borne pathogens and spoilage microorganisms in meat. Interest on the use of fermentation techniques in meat has been revived since there is now a restriction on the use of chemical additives for preservation of meat and meat products. Fermentation is a natural biological process that can be easily adopted in developing countries where refrigeration facilities are lacking Meat fermentation is a complex biological phenomenon accelerated by the desirable action of certain microorganisms in the presence of a great variety of competing or synergistically acting species mainly acquired from meat. Dry-cured sausages are one of the oldest forms of preserving meat and are typical of Mediterranean countries with a dry climate (Spain, Italy, France, Portugal and Turkey). In contrast, smoke cured sausages, or cooked sausages prevail in countries with a colder climate. In general, the qualitative characteristics of naturally fermented sausages are known to be largely dependent on the quality of the ingredients and raw materials, the specific conditions of the processing and ripening, and the composition of the microbial population. Since different genera and species, and even strains, have been shown to significantly affect the sensory traits of fermented sausages, the microbial ecology of fermented sausages has become of increasing interest over the last few decades.

Microorganisms used in Meat Fermentation

Meat is mostly subjected to deterioration by the growth of several microorganisms. While these spoiling microorganisms are not acceptable in raw meat and meat products, certain types of fermentative microorganisms, especially the LAB, either exist in raw meat naturally (natural micro flora) or added by producers, are used for the production of fermented meat products. These desirable microorganisms added to the meat dough are called as starter cultures and they can be single species or the mix of certain microorganisms.

The microorganisms of importance during fermentation and maturation of fermented sausages are Gram positive and rod shaped, belonging to the genera Lactobacillus, Micrococcus and Staphylococcus In some fermented sausages, particularly those produced in France, Spain and Italy, the sensory properties of the products are also influenced by the development of the surface flora, consisting of moulds and yeasts The surface flora of many air-dried dry sausages consists mainly of moulds of the genus. Penicillum and protects the products and gives them typical aromas.

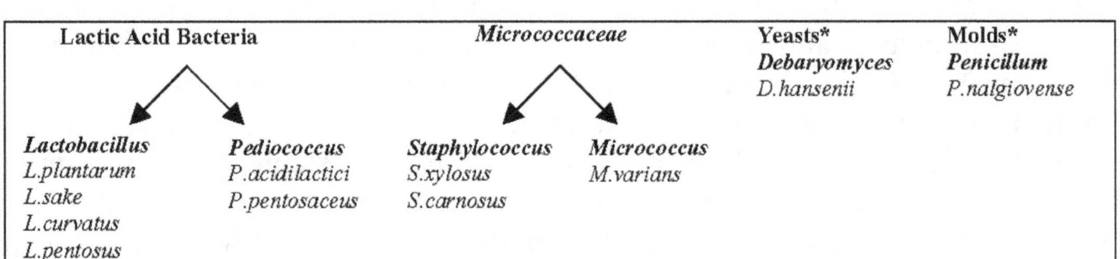

Microorganisms used in starter culture preparations.

Most fermented foods, including the major products that are common in the western world, as well as many of those from other sources that are less well characterized, are dependent on LAB to mediate the fermentation process LAB enhance the physico-chemical properties of sausages and restrict the growth of some undesirable microorganisms. However, LAB have also been reported as major spoilage microorganisms in meat products. These types of microorganisms may cause slime and sour odour formation in sausages.

The broadly used microorganisms in starter culture preparations are given in figure. All LAB produce lactic acid from hexose and since they lack functional heme linked electron transport chains and a functional Krebs cycle; they obtain energy via substrate level phosphorylation. The lactic acid produced may be L(+) or, less frequently, D(-) or a mixture of both. LAB are generally mesophilic but can grow at temperatures as low as 5°C or as high as 45°C. Similarly, while the majority of strains grow at pH 4.0-4.5, some are active at pH 9.6 and others at pH 3.2. Strains are generally weakly proteolytic and lipolytic and require preformed amino acids, purine and pyrimidine bases and group B vitamins for growth.

Micrococcaceae species are used to enrich fermentative microorganisms during aging of the fermented meat products in order to enhance the colour stability of the cured-meat and prevent rancidity. The activity of this microbial group prevents the growth of spoiling microorganisms, decreases processing time and contributes to flavour development It is generally accepted that micrococci participate in desirable reactions occurring during the ripening of dry fermented sausages, such as colour stabilisation, decomposition of peroxides, proteolysis and lipolysis. Some studies clearly show that the aroma of fermented meat products can be modulated by the presence of Staphylococcus spp. The relative importance of Micrococcaceae and lactic acid bacteria in aroma development has recently been reviewed and it is reported that S.xylosus and S.carnosus are able to produce esters and other important aromatic components from amino acids. These strains also prevent the formation of off-flavours, mainly by their high nitrate reductase and catalase activities.

The microorganisms used as starter culture should have some special effects on certain quality characteristics of fermented meat products. The colour of the product is affected by the drop in pH levels that is because the activity of LAB and likewise nitrate reduction, O_2 utilization and H_2O_2 decomposition effects of Micrococaceae influence the colour. Also the colour of the products in which moulds and yeasts are used as starter culture is affected by H_2O_2 decomposition and O_2 utilization activities of these microorganisms as well. Aroma is another important quality feature of fermented meat products. It is directly affected by the starter cultures. LAB affect the aroma with producing acid while Micrococaceae, moulds and yeasts have more efficiency on this property by decomposition of proteins, fats and delaying action on rancidity. LAB are the only genera responsible for the texture of the product with pH decreasing effect. However, in some products such as Hungarian Salami, moulds and/or yeasts are the microorganisms responsible for appearance, prevention from O_2 and light and they protect the product against drying. The textural or structural stability in market is also one of the main quality indicators for fermented meat products. It's important that the product can keep its quality characteristics stable until consumption. Drop in pH, nitrate reduction and prevention of spoilage microorganisms in product are the main effects of LAB and Micrococaceae on the market stability of product. Finally, nitrite decomposition effects of LAB and Micrococaceae are helpful for decreasing the level of chemical residues in the product It is recommended that starter cultures for production of biogenic amines be tested in model media. Starter cultures can influence production of biogenic amines in dry fermented sausages by two ways: indirectly, by its ability to depress the growth of microorganisms with decarboxylase activity and, directly by its own decarboxylase activity.

Plant-Based Fermented Foods

The fermentation of plant material is an ancient preservation method, which likely originated in Asia. Nowadays, the traditional Korean product, kimchi, is the most popular fermented vegetable food in many Asian countries. The most common fermented vegetable products in Europe and North America (Canada and the United States) are sauerkraut, cucumbers, and olives. Before the advent of modern food preservation technologies such as refrigeration, controlled atmosphere storage, and freezing, the consumption of fermented vegetable products ensured continuous access to essential vitamins including vitamin C, minerals, and dietary fibers for people in Northern climates with just one annual growth season.

Kimchi

Introduction

Kimchi is a traditional Korean fermented vegetable product, which is most commonly prepared with Chinese cabbage as the main ingredient in combination with many minor ingredients such as red pepper powder, ginger, garlic, radish, and fish sauce. Kimchi preparation methods differ depending on the variety of kimchi and the ingredients used, as well as on whether the production takes place domestically or industrially. Also, the types of kimchi differ from region to region because of differences in harvest and weather conditions. The final flavor profile is determined by the ingredients, condiments, the addition of salt and spices, and fermentation. According to a review by researchers, approximately 190 different types of kimchi are produced in Korean households. In general, kimchi tastes best when it is fermented for 2 to 3 weeks at 2 °C to 7 °C, to reach a pH level of 4.2 and a titratable acidity of 0.6% to 0.8%.

Fermentation Process

The principal process consists of pre-treatment, brining, blending of ingredients, and fermentation steps. For example, the preparation method for Baechu kimchi (cut Chinese cabbage kimchi) is as follows: pre-treated Chinese cabbage is cut into cubes of 3 to 5 cm, brined in a salt solution (8–15% w/v) for 2 to 7 hours, rinsed with fresh water, and drained. In a separate process, sliced Oriental radish, carrots, green onion, onion, chopped garlic, ginger, red pepper, salt-pickled seafood, other minor ingredients, and dry salt are combined to make a pre-mixture according to the specific recipe. This pre-mixture is blended well with the treated cabbage. This mixture is then placed in a fermentation vessel sealed with a lid and fermented either for 1 to 3 weeks at low temperatures of 2 °C to 10 °C or for 1 to 3 days at 20 °C to 25 °C. After fermentation, kimchi is usually stored at temperatures between 0 °C and 8 °C and served cold.

The fermentation occurs mainly due to LAB and yeast strains, which are naturally present in the diverse microflora of the raw material. The fermentation of kimchi is initiated by Leuco. mesenteroides under anaerobic conditions. This organism differs from other LAB species in that it can tolerate fairly high concentrations of salt and sugar. Leuco. mesenteroides initiates growth in vegetables more rapidly over a wide range of temperatures and salt concentrations than any other LAB. Its hetero-fermentative metabolism produces carbon dioxide and acids which rapidly lower the pH and inhibit the development of undesirable microorganisms. The carbon dioxide produced replaces the oxygen, making the environment anaerobic and suitable for the growth of subsequent species of lactobacillus. Removal of oxygen also helps to preserve the color of fermented vegetables and stabilizes the ascorbic acid (vitamin C) that is present in the vegetables. As the pH drops to between 4.6 and 4.9, Leuco. mesenteroides becomes relatively inhibited, but the fermentation continues with other LAB such as Streptococcus (now classified as Enterococcus) faecalis, Lb. brevis, Pediococcus cerevisiae, and Lb. plantarum. Researchers analysed the kimchi microflora using denaturing gradient gel electrophoresis, a 16S rRNA gene–based culture-independent molecular technique, and found that W. confusa, Leuco. citreum, Lb. sakei, and Lb. curvatus were the main microorganisms responsible for kimchi fermentation. Other techniques, which have been used to analyze the microbial community dynamics during kimchi fermentation, include genome-probing microarray for quantitative, high throughput monitoring, polymerase chain reaction–enzyme-linked immunosorbent assay for the screening of many bacterial isolates from fermented vegetables and sodium dodecyl sulfate–polyacrylamide gel electrophoresis profiles of whole cell proteins for the identification of LAB in kimchi.

The most important factors that affect kimchi fermentation are the endogenous plant microorganisms, salt concentration, fermentable carbohydrates, other available nutrients, presence of inhibitory compounds as well as oxygen, pH, and temperature. The effect of salt concentration, temperature and pH is mostly due to the influence of these factors on the rate and extent of lactic acid fermentation. The optimum fermentation time to obtain the most acceptable quality of fermented kimchi with different salt concentration and temperature combinations has been investigated by researchers. They found that the optimum period of fermentation for Baechu kimchi is 1 to 2 days at 30 °C with a 3.5% salt concentration, and 2 days with a 5% salt concentration and the same temperature. At 14 °C, 5 to 12 days are required if combined with 3.5% salt, and this increases to 10 to 18 days if 5.0% salt is used.

The use of well-defined starter cultures is of considerable commercial interest as it can improve the control and predictability of the fermentation process. Researchers isolated the dominant Leuco. citreum IH22 in their polyphasic study of the LAB community in kimchi fermenting at 15 °C and successfully used this strain to obtain a faster and more consistent fermentation to yield a product with better organoleptic properties than the control kimchi.

Microbiological Characteristics

The kimchi fermentation pattern is similar to that found in other vegetable lactic acid fermentations; however, due to the many types of kimchi products, comparatively greater variance is found in the associated LAB flora than in other products. Also, because the raw ingredients for kimchi are traditionally not blanched before use, the epiphytic LAB as well as the rest of the other microflora associated with the plant material in the field will be retained to possibly influence the fermentation.

As with other vegetable fermentations, the brining step is very important for kimchi fermentation. Brining extracts the water and nutrients from the raw materials by osmotic activity and suppresses the growth of some undesirable bacteria that may spoil the kimchi. At the same time, it makes conditions relatively favorable for LAB by increasing the salt content in the Chinese cabbage or radish. The LAB count increases approximately fourfold after brining of the Chinese cabbage. Also, after the brining and rinsing/washing treatments, the counts of bacteria, yeasts, and molds are greatly decreased.

The fermentation containers are usually covered with lids to provide anaerobic conditions and thus minimize the growth of aerobic microorganisms. Leuco. mesenteroides is the predominant

LAB in the early fermentation stages but, because of the decrease in pH, is gradually replaced by other LAB such as E. faecalis, Lb. brevis, P. cerevisiae, and Lb. plantarum. There is a considerable overlap of LAB species. The growth of each species depends on its initial number in the raw material (Chinese cabbage and other ingredients), the concentrations of salt and sugar, the absence of oxygen, and the fermentation temperature. Lb. plantarum is commonly present in the greatest numbers after the initial fermentation and results in maximum acidity at the later stages. Aerobic yeasts and molds may appear on the surface of improperly covered vegetables at the later fermentation stages. The appearance of undesirable and spoiled kimchi, which is characterized by off-flavors and softened texture, is possibly due to excessive aerobic growth of molds and film-forming yeasts. Softening and excessive acidification of kimchi at the overripening stages during fermentation and storage are the most serious problems. It should be pointed out that once prepared, kimchi is ready-to-eat at any stage of the fermentation depending on the individual consumer's preferential taste. Because there is no sterilizing process during kimchi processing, microbial quality and safety are very important issues for the shelf life of kimchi.

Sauerkraut

Characteristics of LAB in Sauerkraut Fermentation

The LAB that dominate Western vegetable fermentations, such as sauerkraut, mainly belong to the genera Lactobacillus, Leuconostoc, and Pediococcus. Homofermentative strains of lactobacilli produce 85% lactic acid from glucose, whereas heterofermentative strains produce lactic acid, carbon dioxide, ethanol, and acetic acid in equimolar amounts. The strains of Lb. sakei and Lb. curvatus are phenotypically closely related, and they are considered beneficial in sauerkraut fermentation.

Previous studies using traditional biochemical identification methods to study the ecology of commercial sauerkraut fermentations revealed that the obligate heterofermenter, Leuco. mesenteroides; the facultative homofermenter, Lb. plantarum; the homofermenter, Ped. pentosaceus; and the obligate heterofermenter, Lb. brevis were the primary microorganisms responsible for fermentation. However, as previously mentioned, using DNA fingerprinting techniques, microbiologists identified more species of LAB to be present in sauerkraut fermentation including Leuco. citreum, Leuco. argentinum, Leuco. fallax, Lb. paraplantarum, Lb. coryniformis, and Weissella sp.

In addition to the desirable LAB, there are undesirable microorganisms present on cabbage which can interfere with the sauerkraut production process. The quality of the final product depends largely on how well the undesirable organisms are controlled during the fermentation process. Some spoilage organisms use the protein as an energy source, producing unpleasant odors and flavors.

Fermentation Process

Shredded cabbage is placed in a container and salt is added. Mechanical pressure can be applied to expel the cabbage juice, which contains fermentable sugars and other nutrients suitable for LAB activity. The heterofermentative gas-producing cocci (Leuco. mesenteroides) start the initial fermentation. When the acidity (lactic acid) reaches 0.25% to 0.3%, the activity of this group of LAB slows down, although their enzymes continue to function. The fermentation is subsequently continued by the lactobacilli (Lb. plantarum, Lb. cucumeris, and Lb. brevis) until the acidity level increases to 1.5% to 2%. The high salt concentration and low temperature inhibit these bacteria to some extent. Finally, Lb. pentoaceticus continues the fermentation, bringing the acidity to 2% to 2.5% to complete the fermentation.

The end products from the sauerkraut fermentation are lactic acid along with smaller amounts of acetic and propionic acids, a mixture of gases (of which carbon dioxide is the principal gas), small amounts of alcohol, and a mixture of aromatic esters. The acids, in combination with alcohol, form esters which contribute to the characteristic flavor of sauerkraut. The acidity helps in controlling the growth of spoilage organisms and contributes to the extended shelf life of the product.

Temperature and Salt

Temperature control is one of the most important factors in the sauerkraut production process. The optimum temperature for sauerkraut fermentation is approximately 21 °C. A temperature range of 18 °C to 22 °C is most desirable for initiating fermentation due to the growth and metabolism of Leuco. mesenteroides. In contrast, temperatures higher than 22 °C favor the growth of Lactobacillus species.

Salt also plays an important role in initiating the sauerkraut production process and affects the quality of the final product. Salt is added to withdraw juice from the cabbage, thus creating a more favourable environment for the development of the desired LAB flora. Generally, salt is added to a final concentration of 2.0% to 2.5%. At this concentration, lactobacilli are slightly inhibited, but cocci are not affected. It is essential to use pure sodium chloride because salts with added alkali may neutralize the acid.

Starter Cultures

Instead of relying on the endogenous LAB flora and inherent variability between different harvest areas, cultivars, processing environments, etc., the use of starter cultures can improve the fermentation process to ensure the safety and health aspects of the products. More importantly, the use of starter cultures results in a more controllable and consistent quality sauerkraut and helps to speed up the fermentation process. Starter cultures also helps ensure that the correct level of acidity is reached in the fermented product such that undesirable (Gram-negative) microorganisms are inhibited. It is possible to add starter culture such as Lactococcus lactis without adverse effects on final product quality. These starters only survive for a short time but initiate the acidification process; therefore, they do not disturb the natural sequence of microorganisms. The racemase-deficient Lactobacillus bavaricus strains are also used as starter cultures in sauerkraut. Scientists successfully used Leuco. mesenteroides with or without malo-lactic activity to ferment sauerkraut. However, although the addition of Leuco. mesenteroides in the early stages of fermentation can provide attractive flavors to the final product, the sequence of subsequent LAB growth may become altered to possibly result in a final product that is incompletely fermented. The correct sequence of organisms is essential in achieving a stable product with typical sauerkraut flavor and aroma. It should be noted that the juice from previous sauerkraut fermentations was traditionally used as a starter culture for subsequent fermentations. The efficacy of using old spent juice depends largely on the types of organisms present in the juice and its acidity.

Spoilage and Defects in the Sauerkraut Process

Most spoilage of sauerkraut is due to aerobic soil microorganisms which break down the protein to produce undesirable flavor and texture changes. The growth of these aerobes can be controlled by the fermentation rate and control of temperature and salt concentrations.

Whenever the normal sequence of bacterial growth is altered, it usually results in a soft product. Lactobacillus sp. seem to have a greater ability than the cocci to break down cabbage tissue, thereby causing the softening. High fermentation temperatures and reduced salt concentrations favor the growth of Lactobacillus sp. If the salt concentration is too low initially, the Lactobacillus sp. will grow rapidly at the beginning and affect the normal sequence of fermentation.

Several factors can favor the growth of spoilage organisms. For example, an uneven distribution of salt and an insufficient level of cabbage juice to cover the sauerkraut during the fermentation can allow undesirable aerobic bacteria and yeasts to grow on the surface of the sauerkraut, causing production of off flavors and discoloration. In addition, if the fermentation temperature is too high, growth of undesirable pigmented microorganisms may result in the development of dark spots.

Pink sauerkraut is caused by a group of yeasts which produce an intense red pigment in the juice and on the surface of the cabbage if the salt is unevenly distributed or added to an excessive concentration, allowing yeast to grow while inhibiting the LAB. Researchers isolated and identified two spoilage yeasts, Saccharomyces bayanus and Saccharomyces unisporus, from spoiled fermented vegetables. In addition, the shelf life of unpasteurized fermented vegetables is influenced by the presence of fermentative yeasts which may grow during storage if the sugar catabolism by LAB is incomplete.

Comparison of Kimchi and Sauerkraut

Although the Western and Eastern fermented vegetable products seem to share many commonalities, there are some differences. For example, the preferred end point of fermentation is different for kimchi and sauerkraut. In the Oriental tradition, the best-tasting kimchi is attained before overgrowth of Lb. brevis and Lb. plantarum happens, with an optimal product pH of 4.5. The overgrowth of Lb. brevis and Lb. plantarum affects the quality of kimchi due to their acidification of product pH values to lower than 4.5. In contrast, sauerkraut production depends on these organisms and the fermentation is specifically manipulated to promote their growth by using elective salt and temperature combinations. The optimal range of salt concentration of sauerkraut is 0.7% to 3.0%, whereas it is 3.0% to 5.0% for kimchi.

Several strains of bacteriocin-producing microorganisms have been isolated from kimchi and sauerkraut. For example, Enterococcus faecium strains found in kimchi have been reported to have a broad spectrum of bacteriocin activities, and several Lactobacillus spp. have also been shown to have an antimicrobial effect. The organic acids, bacteriocin(s) produced during fermentation, and the antimicrobial activity of the ingredients together contribute to the control of undesirable microflora, including pathogenic microorganisms, without costly preservation treatments and packaging.

Biopreservation of Vegetable Products

The idea of using biopreservation with innocuous commensal bacteria to inhibit foodborne microbial pathogens and spoilage organisms is not new but has mainly been explored in meat and fish products. This purposeful use of microorganisms and their metabolites aims to extend the shelf life and improve the safety of foods. It is, in particular, biopreservation by use of active LAB cultures and their metabolites including bacteriocins that has gained considerable interest both in terms of their antibacterial and antifungal effects.

As fresh-cut, ready-to-eat fruit and vegetable products are becoming more popular with consumers, the food industry needs to seek out ways of managing the associated microbial hazards and quality problems. Biopreservation with LAB of plant origin may present an attractive alternative to chemical washes and preservatives. It is generally recognized that LAB best suitable for biopreservation should originate from the same type of food material as they are intended for use. In a search for plant LABs with biopreservative properties, Trias and colleagues isolated 523 LAB strains from more than 700 samples of fruits and vegetables. After an initial screening, they selected 18 antagonistic strains for further characterization. Challenge experiments with Leuco. mesenteroides, Leuco. citreum, Lb. plantarum, Lact. lactis, and Weisella cibaria on apples and lettuce showed that four Leuconostoc sp. strains completely inhibited Listeria monocytogenes and reduced the growth of Escherichia coli and Salmonella typhimurium. Scientists investigated the antifungal properties of plant-derived W. confusa, W. cibaria, Lb. plantarum, and Ped. pentosaceus strains and found activity against a range of spoilage fungi.

Fermented Drinks

The global functional beverage market is a growing sector of the food industry as modern health-conscious consumers show an increasing desire for foods that can improve well-being and reduce the risk of disease. Fermented milks, especially yoghurt-style products, are the most popular functional beverages with kefir in Western Europe and North America and ymer in Denmark being good examples. Notably, the global functional food and drink market increased 1.5 fold between 2003 and 2010, and is expected to grow a further 22.8% between 2010 and 2014 to be worth €21.7 billion, with other estimates predicting the market will reach €65 billion by the year 2016. Dairy-based produce account for approximately 43% of the functional beverage market, and is mainly comprised of fermented products. It is also intriguing to note that a number of food companies that have been under pressure, due to the poor public perception regarding the 'healthiness' of the foods they produce, are now focusing on developing such functional products.

The global functional beverage market is a growing sector of the food industry as modern health-conscious consumers show an increasing desire for foods that can improve well-being and reduce the risk of disease. Fermented milks, especially yoghurt-style products, are the most popular functional beverages with kefir in Western Europe and North America and ymer in Denmark being good examples. Notably, the global functional food and drink market increased 1.5 fold between 2003 and 2010, and is expected to grow a further 22.8% between 2010 and 2014 to be worth €21.7 billion, with other estimates predicting the market will reach €65 billion by the year 2016. Dairy-based produce account for approximately 43% of the functional beverage market, and is mainly comprised of fermented products. It is also intriguing to note that a number of food companies that have been under pressure, due to the poor public perception regarding the 'healthiness' of the foods they produce, are now focusing on developing such functional products.

Natural Fermented Beverages: Sources and Microbial Composition

Naturally Fermented Milks

The yoghurt and fermented milks market is currently worth €46 billion, with North America, Europe and Asia accounting for 77% of the market. Many communities across the world produce naturally fermented milks with many of these products being of a yoghurt-style consistency. Fermented milk products can be made with milk (or skimmed milk) from various sources, including cow, camel, goat, sheep, yak and even coconut, milk, and can be either pasteurised or unpasteurised. They can be produced through the use of defined starter cultures, back-slopping or allowed to ferment naturally. Although fermented milk beverages are predominantly composed of lactic acid bacteria (LAB), the exact microbial content may vary depending on the source of milk, treatment of the milk (e.g. pasteurisation), use of starters, the nature of the local environmental microbes present, temperatures, hygiene, the type and treatment of containers used and the length of fermentation. Many artisanal fermented milk beverages are produced as a result of back-slopping, whereby a small portion of already-fermented milk is used to begin a new fermentation. In this way, cultures from the LAB naturally present in the raw milk are passed from household to household and between generations. While the consumption of spontaneously fermented milk is common to many different regions, the exact microbial differences between these products have not been ascertained. Table lists a number of the most popular and best-studied fermented beverages from around the world, along with information with respect to their corresponding microbial compositions. From this, the domination of milk-based beverages fermented by LAB, mainly Leuconostoc, lactobacilli and lactococci, is clear. Fermentation in colder climates promotes the growth of mesophilic bacteria such as Lactococcus and Leuconostoc, whereas beverages produced at higher temperatures usually have greater counts of thermophillic bacteria such as Lactobacillus and Streptococcus. The contributions of slime-producing species or acetic acid producing species, generally present at low abundance relative to Lactobacillus or Lactococcus species, vary depending on abundance. There may also be significant numbers of coliforms present, depending on the level of hygiene employed during preparation, with high levels having been noted in some African beverages. The quantity and types of yeasts involved can vary greatly, but Candida and Saccharomyces are the species most commonly detected.

Of the many fermented milk beverages, kefir, a drink that originated with shepherds in the Caucasian mountains has been a notable success, gaining worldwide popularity, with the market now worth €78.7 million in North America alone. The microorganisms responsible for the fermentation are actually a symbiotic combination of bacteria and yeast, bound within a polysaccharide matrix, known as kefir 'grains'. Koumiss, sometimes known as airag, is a popular beverage of nomadic cattle breeders in Asia and some regions of Russia. This beverage is similar to kefir, but there is no solid inoculation matrix, and this milk is fermented by back-slopping or by allowing the milk to ferment naturally, and has been reported to contain fewer lactococci. Shubat is a fermented camels milk popular in Asia, also believed to have healing properties. In Africa, fermented milk beverages are quite popular, where the art of making fermented products is passed down through generations. Examples of such beverages include amasi from Zimbabwe, kivuguto from Rwanda, suusac from Kenya, nyarmie from Ghana and rob and garris from Sudan. Considering that most of these are derived from the spontaneous fermentation of milk by its innate microbiota, it is likely that the fermented milks, although known by different names, are actually quite similar, and can be, in combination, referred to as naturally fermented milk (NFM). Nonetheless, accurate categorization remains difficult in the absence of more detailed microbiological and biochemical analyses. Also, in many countries yoghurts are diluted with water to form drinkable fermented milk, such as doogh, ayran, chaas and lassi, with the resulting microbial composition generally being similar to that of yoghurt.

Non-Dairy Fermented Beverages

Another important class of fermented beverages are those made from cereals, which are popular in tropical regions and on the continent of Africa in particular. As with many milk-based products, the natural microbial component is used to ferment grains including maize, millet, barley, oats, rye, wheat, rice or sorghum. The grains are often heated, mashed and sometimes filtered. Back-slopping is again quite common, but the microbial populations responsible for the fermentation of these beverages are not as well characterised.

Boza, consumed in Bulgaria and Turkey, is generated through the fermentation of a variety of cereals including barley, oats, rye, millet, maize, wheat or rice, with the specific composition affecting the viscosity, fermentability and content of the final beverage. The cereal is boiled and filtered, a carbohydrate source is added, and the mixture can be left to ferment independently or with the use of back-slop. Boza has yet to be commercialised and studies have revealed that the microbial population varies. The function of the yeast present, which are only sometimes detected, remains unknown. Of several combinations, it has been suggested that fermentation by S. cerevisiae, Leuconostoc mesenteroides and Lactobacillus confusus produce the most palatable beverage.

Togwa, a sweet and sour, non-alcoholic beverage, is one of the better studied African cereal beverages. Produced from the flour of maize, sorghum and finger millet and, sometimes, cassava root, the chosen substrates are boiled, cooled and fermented for approximately 12 hours to form a porridge, which is then diluted to drink. Mahewu is similar in that maize or sorghum meal is fermented with millet or sorghum malt, and is available commercially. Bushera is generally prepared from germinated or non-germinated sorghum grains, and fermented for 1-6 days. These beverages are often used to wean children, and as a high-energy diet supplement. Koko sour water is the fermented liquid water created in the production of the fermented porridge, koko. This contains a high portion of LAB and is used by locals to treat stomach aches and as a refreshing beverage.

Kvass is a fermented rye bread beverage common in Russia, which has seen much commercial success. The beverage can have a sparkling, sweet or sour, rye bread flavour. Its alcohol content, though usually low, can vary, and has been suggested as a contributor to alcoholism. Amazake is a sweet fermented rice beverage that is the non-alcoholic precursor to sake, produced in Japan. Steamed rice is mixed with rice-koji (Aspergillus-mycelia and rice) and water, and is heated to 55-60 °C for 15-18 hours. Enzymes break down the rice and form glucose content of approximately 20%. Amazake is highly nutritious and is consumed for its purported health benefits. Pozol, which is common to south-eastern Mexico, has quite a different method of production, in that maize grains are heat-treated in an acid solution, ground and shaped into dough balls. These are then wrapped in banana leaves and fermented for 2-7 days, after which they are re-suspended in water and consumed as beverages. Pozol is composed of a variety of microorganisms including LAB, non-LAB, yeasts and other fungi.

In addition to milk and cereal-based fermentations, there are also other forms of fermented beverages. One example is kombucha, which is a fermented sweetened tea that was originally popular in China but is now enjoyed worldwide, and is set to be worth €363 million by 2015 in North America. It is fermented by a symbiotic mixture of bacteria (typically acetic acid bacteria, with small quantities of LAB) and yeast, which are embedded within a cellulosic matrix that floats above the fermentate, similar to the mother cultures of vinegar. Due to the high acid content (as low as pH2), the functionality of kombucha is predominantly due to its physiochemical properties. As a result of the tea content, it also contains a number of phenols and vitamins. Water kefir is similar in concept to milk kefir in that it is fermented by a symbiosis of bacteria and yeast contained within grains. However, these grains are composed of dextran, are translucent and crystal-like in appearance, and are thought to have originated in Mexico where they formed as hard granules fermented from sap on the pads of the Opuntia cactus. They ferment sweetened water, to which figs and lemon are traditionally added for additional flavour and nutrients. The composition of water kefir can vary, but is known to contain LAB, including Lactobacillus, and Bifidobacterium. Hardaliye is a non-alcoholic, Turkish, fermented beverage made from red grapes, black mustard seeds, cherry leaf and benzoic acid. Ingredients are pressed and fermented for 5-10 days at room temperature. Again, the microbial population has been reported to be predominantly composed of Lactobacillus and unknown fungal components, and this beverage is thought to have antioxidant properties.

Health Benefits

Originally devised as a means of food preservation, over time many beverages, such as kefir and koumiss, became popular due to their reputed abilities to improve gastrointestinal health. However, most of the traditional fermented beverages are poorly studied, with unsubstantiated claims linking them to positive effects on human health. Ideally, any such beverage making health claims should be backed by credible scientific evidence in the form of randomised, controlled and replicated human intervention trials. This form of evidence is rare for these beverages (and particularly so for non-dairy forms), and the generation of such data is an expensive and unappealing prospect for industry, but nonetheless remains a critical area for proof-of-concept and future research. Despite this, however, there is still a perception that many of these beverages are "healthy", particularly in societies where the beverage is steeped in local tradition, which in turn contributes to their market potential and justifies investing in related research.

For many of the fermented beverages, it is the strong association between the microbial content and improvement of gastrointestinal health that is thought to be responsible for perceived health outcomes. While it is sometimes unclear what functional characteristics traditional beverages confer beyond the basic nutrition of the raw unfermented ingredients, there is evidence that some fermented beverages provide beneficial effects through direct microbial/probiotic action and indirectly via the production of metabolites and breakdown of complex proteins. Nonetheless, natural fermented milks have been shown to have antihypertensive effects, enhance systemic immunity, lower cholesterol and to help lower blood pressure. In recent human trials, they have been shown to aid in the treatment of IBS and to help alleviate constipation. Additionally, they have been shown to have modulatory effects on the brain, and demonstrate anti-cancer potential. Of the traditional-style beverages, kefir specifically has been shown to positively impact the gastrointestinal tract, stimulate the immune system, and have anti-inflammatory and anti-carcinogenic effects, albeit not through clinical trials. Lactic fermented milks often contain compounds not present in regular milk, such as exopolysaccharides, e.g. kefiran in kefir, and natural enrichments, including increased vitamin (e.g. B12 and K2), folate and riboflavin content. Furthermore, fermented dairy products usually possess β-galactosidase activity and reduced lactose content compared to milk, making them potentially suitable for those suffering from lactose intolerance. Fermented produce can also be a source of bioactive peptides, released through fermentation by proteolytic cultures, and have been linked with many potential health benefits including digestive, endocrine, cardiovascular, immune and nervous system affects.

The occurrence of organic acids, which lower the pH of the beverages, may also confer health benefits. Indeed, the presence of glucuronic acid, one of the primary metabolites in kombucha, is believed to improve detoxification by binding toxin molecules and aiding excretion through the kidneys and it is this acidic composition that is most associated with the reputed health properties of kombucha, rather than a microbial-gut interaction. Kombucha also contains increased B vitamins and folic acid in addition to a number of healthy components, such as phenols, naturally present in tea. Acid content, in conjunction with antimicrobials often produced by bacteria, could result in the beverage possessing therapeutic, antimicrobial properties.

Fermented cereals can also contain a high mineral content, and generally have a lower fat percentage than their dairy-based counterparts, but grains are generally lacking in essential amino acids. These forms of beverages can also naturally provide plant-based functional components, such as fibre, vitamins, minerals, flavonoids and phenolic compounds, which can effect oxidative stress, inflammation, hyperglycemia and carcingenesis. As previously mentioned, fermented foods are particularly common in Africa, where palates are accustomed to sour foods. Providing a safe, fermented cereal beverage with reliable probiotic cultures could help reduce diarrhoea and malnutrition caused by contaminated traditional beverages used in weaning children, and help reduce fatalities and improve well-being.

Despite the need for definitive studies demonstrating direct health benefits on consumers, in vitro and animal studies give reason to be optimistic. In many cultures, alleged health benefits are the reason for consumption, and if there are indeed health benefits to be gained from consuming fermented beverages, it is most likely the result of a synergistic effect between substrates, delicate microbial content and microbial end products, the relationship between which should become clearer with further research.

Beyond Physiochemical Advantages: From Microbial Content to Functionality

Molecular-Based Microbial Characterisation

Despite health claims linked to the microbial composition of fermented beverages, there is a considerable lack of analyses relating to the microorganisms present and the quantities in which they exist in such beverages. In order to address this, it is necessary for the application of unbiased, standardised techniques to assess beverages from different geographical regions, and to reach a consensus on the definition of microorganisms which constitute part of any particular beverage. Since many such beverages are naturally fermented, and thus subject to environmental influences, their microbiota can differ significantly, but the application of reliable technology can help definitively identify a core population (or lack thereof), responsible for characteristic traits of the beverage in question. While some molecular-based, microbial characterisation of these beverages has taken place, most studies have relied on low-throughput approaches, employing techniques such as DGGE, which can only assess 1-2% of a population.

Moving forward, the availability of molecular technologies such as culture-independent, high-throughput, sequencing-based microbial analyses, metabolomics and bioinformatics will prove particularly useful, and will provide a more accurate picture of these populations, surmounting problems associated with relying on phenotypic-based approaches. In-depth molecular studies have the potential to be particularly useful when carrying out analyses across different beverages with a view to attributing specific desirable or non-desirable sensory and organoleptic characteristics with specific microorganisms present. Such approaches will also ultimately facilitate accurate species identification, leading to novel starter design, and the development of beverages with different and complex flavour profiles. It will also be possible to more effectively monitor the change of proportions of different species throughout fermentation and storage. Future studies will also shed light on the nature of the symbiosis of such beverages, which is so complex that in vitro synthesis of kefir grains has yet to be replicated. Currently, commercial kefir is produced by defined starters, with probiotic strains added to some products to boost reputed health claims.

Health-Promoting Microbes

It is widely believed that the primary reason for the functionality of these beverages is due to the presence of specific live microorganisms. To the consumer, health claims are more important than nutritional claims, so there has and will be a desire to augment the health-promoting potential of these beverages through the addition of certified probiotics. The probiotic market was worth €15.7 billion in 2010, and is expected to increase to €22.6 billion by 2015. The WHO/FAO defines probiotics as "live microorganisms, which when administered in adequate amounts confer a health benefit on the host", and the probiotic sector is the largest component of the functional food market. The physiology of certain strains of lactobacilli and bifidobacteria make them well-suited to both the gastrointestinal and milk environments, and thus lactic acid bacteria and bifidobacteria are the most studied and utilised probiotic organisms. It is generally considered that a minimum of 10^9 cells per daily dose are required for probiotics to be effective. Within the EU, the term "probiotic" is now considered a health claim, with strict criteria surrounding its use and resulting in many applications submitted to the European Food Safety Authority (EFSA) being rejected. In Europe, boosting numbers of Lactobacillus and Bifidobacterium in the gut is not deemed to be of sufficient merit to be considered a health benefit; the link must be made to a physiological (e.g. strengthening the immune system or resistance to infections) benefit to the host. Proving such health claims is expensive, and in the midst of unclear definitions and guidelines, industries are currently more likely to develop and market probiotic products in other parts of the world. In situations where probiotic strains are added during fermentation, they must not interact antagonistically with starter strains. This becomes less of an issue if strains are added after fermentation is complete, due to the low metabolic rates at refrigerated temperatures. Additionally, microencapsulation technology may aid in the delivery of probiotic strains by protecting them in non-native environments. In one instance, microencapsulation of Bifidobacterium successfully increased viable numbers in mahewu, without significantly impacting on flavour, suggesting it could be an effective probiotic delivery system.

A health-related role for the yeast in fermented beverages has yet to be elucidated. The volume of studies reporting significant numbers of yeast in traditional fermented beverages indicates their importance in these fermentations. Yeasts in dairy produce generate desirable aromatic compounds, proteolytic and lipolytic activities and can aid bacterial growth by producing amino acids, vitamins and other metabolites, and contribute to the final composition of the product by producing ethanol and carbon dioxide. In particular, studies have demonstrated that yeast can exert a positive effect on the abundance of Lactobacillus in fermented environments, and this might be a key function in such symbioses, as well as preventing the proliferation of undesirable species. While yeast only comprise <0.1% of the gut microbiota, they are 10 times larger than prokaryotes and can thus impede colonisation of pathogenic bacteria. Success has been made in incorporating them in commercial fermented milk products, but excessive gas production during storage can be an issue. Some species of Saccharomyces and Candida yeasts are common to both fermented beverages and the gut microbiota, such as species, and could be investigated with a view to their contribution to fermentations and optimising health-promoting potential. However, to date, Saccharomyces boulardii is the only recognised probiotic yeast.

Rational Design of Starter Cultures

The selection of appropriate starter strains will be key in efforts to accurately reproduce the desirable characteristics of traditional health-promoting beverages for mass production. To faithfully reproduce these beverages and traits, microbes should be sourced from the traditional fermented beverages, given that these microbes have adapted over thousands of years to their respective environments, and are more likely to function at the appropriate pH, salt concentration, temperature etc. For instance, amylolytic digestion of starch could be considered desirable for fermented cereal production, and isolates from boza and pozol have been shown to be capable of this metabolic trait. Such populations also have a history of safe human consumption. Rational strain selection to produce the correct balance of flavour, aroma, texture, acidification, bitterness, speed of fermentation, and the optimum quantity of organic acid, vitamins and minerals is essential, as beverages that are too sour or bitter, or contain too much ethanol, will not meet consumers' approval. Over recent years, genetic tools have become available to engineer and select superior starter strains, but legislation currently hinders their industrial use. The inclusion of strains producing antimicrobials, such as bacteriocins, could serve as natural preservatives and help produce a more natural product, while sequential fermentation with yeast, followed by bacteria, could produce a beverage with the desired physiochemical effects, but without bio-stabilisation issues created by excessive gas production.

As stated above, the natural fermentation of beverages involves many different strains of bacteria, and sometimes, yeast. There is an understandable tendency to keep starter formulations simple but, as traditional beverages show, there are often multiple strains involved, including different species or even microorganisms. From a health perspective, multistrain or multispecies probiotic beverages may provide greater beneficial effects than monostrain cultures. Unfortunately, however, there is a lack of studies assessing the effects of combining several natural strains on the physiochemical and sensory characteristics of milk or other functional beverages. Without such information, it is difficult to accurately reproduce the characteristics of the organic beverage with one produced by a defined combination of starters, to match the flavour and properties of the original beverage. This is crucial when marketing beverages to consumers already familiar with the artisanally produced variant of the product, and if wishing to retain any health-promoting characteristics attributed to the original product.

In spite of the wide range of options available when designing novel health-promoting fermented beverages, there will always be an attraction for healthy foods derived from natural processes. Applying the solid inoculation matrices of traditional fermented beverages to new substrates provides a means of generating new beverages while retaining natural microbial populations. For example, kefir grains have been employed to produce whey and cocoa pulp beverages containing potentially health-promoting strains. Similarly, the cellulosic pellicle of kombucha has been successfully used to ferment milk and other substrates.

Biotechnology and Beverage Development

Expanding technological capabilities, especially ingredient exploration and development, has led to increased functional product innovation. The number of new products with functional claims has been growing by approximately 28% per year. Consumers' willingness to pay a premium price for fortified products is also a key driver for innovation. While most current functional beverages are aimed at the high-income consumer, there is argument to be made that those who would benefit most from fermented beverages are from underdeveloped nations, where such beverages could provide a cost-effective means delivering much-needed nutrition.

Substrate Exploration

The US, Europe and Japan markets account for over 90% of total functional foods, with the majority being functional dairy products. However, non-dairy probiotic delivery has been attracting more attention in recent years, partly due to the success of bio-functional foods and the desire to expand and provide an alternative probiotic choice to conventional dairy-based beverages. Indeed, this market is projected to have an annual growth rate of 15% between 2013 and 2018. Non-dairy probiotic beverages are particularly attractive due to their lack of dairy allergens, low cholesterol content and vegan-friendly status. Furthermore, different substrates can provide different combinations of antioxidants, dietary fibre, minerals and vitamins.

To this end, cereal-based beverages could be marketed in response to consumers' awareness of the benefits of high fibre diets. They contain natural prebiotic traits due to the presence of indigestible fibres and the presence of diacetyl acetic acid aromatic compounds make them palatable, and furthermore, could be cheaper to produce. Oats, a major source of beta-glucan which can reduce LDL-cholesterol, are known to function as a prebiotic by boosting bifidobacteria numbers in the gut and have been investigated with a view to producing synbiotic beverages. Indeed, a fermented oat drink with two Bifidobacterium longum strains was shown to normalise bowel movements in elderly patients. Malt and barley have also been used as beverage substrates, while Emmer, an ancient European cereal has also shown potential as a functional cereal beverage.

There has also been a positive trend towards the consumption soy products, as evident in worldwide soy food sales, which increased from €218 million to almost €2.9 billion between the years 1992 and 2008, and continues to increase. Soy-based beverages contain low cholesterol and low saturated fats, are lactose-free, are rich in iso-flavones and antioxidants, and have been shown to exert beneficial effects on the host. Soy milks are capable of fermentation by probiotic strains and, when fermented by Bifidobacterium and Lactobacillus, have been shown to have a positive impact on the ecosystem of the intestinal tract. Positive consumer attitudes towards soy have encouraged industry to develop probiotic derivatives with several varieties already available commercially.

The utilisation of waste products to generate functional beverages has seen increased interest, with whey being the most prominent example. Whey is a by-product of the cheese industry, which retains 55% of milk nutrients and contains only 0.36% fat, and has the potential for further use in the human diet. In an effort to add value to whey, numerous studies have investigated its fermentation by lactic acid bacteria (Streptococcus and Lactobacillus) to produce a lactic probiotic beverage, and probiotic bacteria have already demonstrated good survival in whey. Prebiotics have also been successfully incorporated, including oligofructose and inulin, and hydrocolloid thickening agents added to improve viscosity and mouthfeel.

One of the most exciting developments is the development of fruit juices, which have been shown to have considerable market value and consumer acceptance. Already considered a healthy food product, fruit juices are often fortified with vitamins and minerals, in addition to having a high nutrient and antioxidant content, and represent a new method of nutrient and probiotic delivery. As an increasing number of studies are demonstrating, sugars naturally present in juices can facilitate the growth of cultures with appealing taste profiles. This is true of tomato, pomegranate, pineapple, orange and cashew-apple juice. These microbes can impact on physiochemical aspects, such as increasing the concentrations of flavanones and carotenoids in orange juice, and have shown good survival rates during storage of the beverages. While the final content of such beverages are quite acidic and best suited to fermentation by probiotic *Lactobacillus* species (*L. casei*, *L. acidophilus*, *L. plantarum*, *L. parachesei* and *L. delbrueckii*), the use of microencapsulation technology could aid in the delivery of other viable probiotic microorganisms. The enrichment of juices with brewer's yeast autolysate before fermentation positively impacts on the nutritional content of the final beverage, raising the feasibility of co-fermentation by the right combination of bacteria and yeast. Examples of commercially available probiotic-containing fruit juices include Biola and Bioprofit. Similar microorganisms have also been shown to successfully ferment various vegetable juices including cabbage, beet, pumpkin, courgette and carrot juices supplemented with prebiotics.

The major challenge with any substrate/culture combination is to overcome the sensory hurdles of sour, acidic fermentates, and produce a palatable beverage that would realistically be consumed regularly to avail of functional benefits. There exists the option of combining fruit or vegetable juices with fermented milks as natural flavourings to overcome undesirable flavours in otherwise promising beverages/products. To this end, the inclusion of sensory panel evaluations provides invaluable information regarding consumer acceptance, especially for non-dairy products which are intrinsically more difficult to sell than their dairy counterparts. The use of direct liquid inoculation systems to include probiotics while avoiding fermentation side-effects has its own problems in ensuring cell viability and stability during storage.

Fermentation Parameters

In addition to the importance and ratio of starter selection, as already described, the fermentation of potentially health-promoting beverages needs to be carefully controlled to achieve stability, sensory and safety standards. Changes in the concentration of sugars and other compounds need to be carefully monitored both during and after fermentation, and is particularly important with respect to the production of ethanol and carbon dioxide. Sensitive techniques, including high performance liquid chromatography (HPLC) and gas chromatographyn (GC), are now routine for such analyses. Antioxidant levels may also be measured by ferric reducing ability of plasma (FRAP) and 2,2-diphenyl-1-picrylhydrazyl (DPPH) assays. pH is obviously crucial to the success of a fermentation and can be lowered to a specified level prior to fermentation to encourage enzymatic activity and prevent contamination. The concentration of oxygen in the brewing environment can also be important depending on the homo- or hetero-fermentative nature of the cultures. Metabolic engineering of fermenting microorganisms may eventually be accepted to boost concentrations of desirable compounds in the final products. Other factors to be considered include the choice of substrates, particularly the types and concentrations of carbohydrates, and the treatment of raw ingredients such as cereals, by homogenisation, for example, can allow for more effective metabolism and release of bioactive peptides. The time and intensity of heat during pasteurisation need to be considered. The addition of certain compounds, such as ascorbic acid and NaFeEDTA, can encourage the release by fermentation of bioactives, such as, zinc and iron, in the final beverage. Conversely, the concentration of phytic acid, an inhibitor of mineral absorption, particularly with regards to cereals, could lower the mineral impact of the final beverage, and the inclusion of phytases might be necessary to ensure or augment health claims. The concentration of bulk starch and other factors will impact the consistency of the final drink, providing either a thin and free-flowing product or a thicker beverage with a smoothie-like consistency. Natural antimicrobials, such as bacteriocins, may be included to act as preservatives and prevent the growth of spoilage organisms. As microbes, particularly yeast, continue to grow following storage, packaging must be able to withstand the pressure generated as a consequence of gas production, with either plastic or glass containers. Additionally, viability during storage needs consideration, particularly for cereal-type beverages that would traditionally be stored at room temperature.

Clearly, there a number of parameters and variations that need to be measured, controlled and experimented with to determine the optimum conditions for fermentation, and proven to be consistent following up-scaling. Automated processes such as controlled nutrient availability and stirring can influence efficiency of the fermentation. Ongoing research into community analysis and fermentation biochemistry will inform future decisions regarding the control of processes for these types of fermentations.

Beverage Enhancement

There are now a variety of enhancements that can be made to both traditional and novel beverages to boost health claims. Prebiotics, including fructooligosaccharides, inulin and galactooligosaccharides, are often added commercially to fermented milks to promote the growth of favourable bacteria, while investigations of other prebiotics such as oligofructose and polydextrose have also yielded positive results. In addition to preventing and treating intestinal-associated diseases, the incorporation of bio-active nutraceuticals such as ω-3 fatty acids, isoflavones and phytosterols in fermented milks also have potential applications. Isoflavones are powerful antioxidants, comparable to vitamin E, while plant-derived phytosterols are cholesterol-lowering agents. Addition of such compounds, however, can be complicated as ω-3 fatty acids are sensitive to light, air and heat, and can cause undesirable flavours in the end product, while isoflavones and phytosterols have poor solubility in water, and might be difficult to incorporate into non-fat solutions. ω-3 fatty acids can now be microencapsulated to hide fishy off-notes.

A variety of vitamins and minerals may be added, including vitamins D, E and C, calcium and magnesium, while fortification of fermented milks with iron was shown to improve the growth of preschool children. Microorganisms can also provide functional metabolites, which has encouraged the screening of ecological niches, such as marine environments, in addition to the previously referred to gut-derived probiotics, for novel nutraceuticals, the likes of which may eventually be incorporated into functional beverages. Finally, it may be necessary to combat or mask resulting undesirable flavours and aromas arising from the addition of functional ingredients, using flavour enhancers.

Chapter 4

Techniques of Food Preservation

All the techniques of food processing that prevent the spoilage of food due to the growth of micro-organisms and slow down the oxidation of fats fall under food preservation. It includes frozen foods, pickling, food sampling and home canning. The chapter closely examines these key concepts of food preservation to provide an extensive understanding of the subject.

Frozen Food

Freezing provides a significant extended shelf life and has been successfully employed for long-term preservation of many foods. Freezing is still one of the most widely used methods of food preservation even though several new technologies, such as high pressure, infrared irradiation, pulsed electric field, and ultrasound, are gaining importance. Freezing processes are continuing to emerge. Freezing changes the physical state of a substance by changing water into ice when energy is removed in the form of cooling below freezing temperature. Usually, the temperature is further reduced to storage level (e.g., -18 °C). The freezing process can be clearly shown by using freezing or cooling curves and phase diagrams.

Mode of Preserving Action

The freezing of foods slows down, but does not stop, the physicochemical and biochemical reactions that govern the deterioration of foods. During storage, there is a slow progressive change in organoleptic quality, which does not become objectionable for some time. The loss of quality of frozen foods depends primarily on storage temperature, length of storage time, and thawing procedure. Microbial growth is completely stopped below −18 °C, and both enzymatic and non-enzymatic changes continue at much slower rates during frozen storage.

The freezing process reduces the random motion and rearrangement of molecules in the matrix. Freezing involves the use of low temperatures, and reactions take place at slower rates as temperature is reduced. The presence of ice and an increase in solute concentration can have significant effects on the reactions and the state of the matrix. The final influence of temperature on chemical reactions due to freezing could be grouped as: (a) normal stability (a temperature decrease results in a slower reaction rate, thus better stability when foods are stored); (b) neutral stability (the temperature has no influence on the reaction rate); or (c) reversed stability (a temperature decrease results in an increased reaction rate). Regardless of the type of aqueous system, concentration during freezing causes the unfrozen portion to undergo marked changes in physicochemical properties such as ionic strength, pH, viscosity, water activity, surface and interfacial tension, and oxidation–reduction potential. It is important to note that oxygen is almost totally expelled from ice crystals as they are formed. Researchers reviewed three types of cell damage due to freezing: osmotic damage, solute-induced damage, and structural damage.

In slow cooling, ice forms slowly in the external cells. If there is sufficient time, water from the cells migrates out by osmotic pressure. This results in cell shrinkage and some membrane damage. This water does not return to the cells on thawing due to the damage to the cell wall, and the consequence is drip loss. The concentration of the solute increases as freezing progresses. Thus, high solute concentrations of the unfrozen matrix, in particular high salt, can cause damage to many polymeric cell components and may kill the cell. This concentration effect is present irrespective of whether freezing is fast or slow. Cryoprotectants, such as sugars, are usually added to the aqueous phase to reduce salt-induced damage. In addition to the concentration effect, the formation of ice within the cell may cause damage to the delicate organelle and membrane structure of the cell. As a consequence, enzyme systems may be released, leading to a variety of effects, including off-flavor production. This can be prevented by blanching, a pre-freezing heat treatment that denatures enzymes.

It is possible to design and control a convenient freezing process through the knowledge of the mechanisms of damage for each particular food item. In general, freezing preservation is far from perfect, and awareness of this fact is needed if techniques are to be developed to overcome known shortcomings and assure that this method remains competitive with the other major methods of food preservation. A strategic quality approach (quality enhancement) may provide a higher success rate for new frozen food products. Product quality improvement and energy reduction or increased process efficiency are major issues related to the freezing process.

Quality of Frozen Foods

Freezing Rate and Quality

Controlling the freezing process, including careful prefreezing preparation and postfreezing storage of the product, is an important aspect of achieving a high-quality product. An important factor in the quality of frozen foods is the freezing rate. Generally, fast freezing produces better quality frozen food than slow freezing, although the reason for this is not as well understood as is sometimes stated. The rate of freezing of plant tissue is important because it determines the size of the ice crystals, cell dehydration, and damage to the cell walls. Ice crystal structure is crucial for the preservation of the quality of frozen products.

In the case of animal tissue, the concentration of salt within the cells is higher than that in the extracellular region. Consequently, freezing will start outside the cells due to the freezing point depression induced by the solute concentration in the cells. As soon as ice appears, the solute concentration rises. This is a characteristic phenomenon of freeze concentration. At some point, osmotic pressure difference causes water to flow through the semipermeable cells to the extracellular region to balance the chemical potentials. This dehydration of the cell is accompanied by shrinkage of the cell, which is not normally lethal. The freezing rate affects this process because rapid freezing results in less cell dehydration (since water has less time to diffuse out of the cell), less breakage of cell walls, and less textural damage. The more rapid the crystallization, the smaller the ice crystals, and the lesser the damage caused by the process of freeze concentration. Consequently, a reverse situation holds for thawing. Slow heating allows equilibrium to be reached as the melted water diffuses back through the cell wall.

In the case of plant tissue, there is evidence that large ice crystals can cause mechanical damage to cell walls in addition to cell dehydration. In agarose gels, large ice crystals (100–300 µm) with increasing interstitial spaces grow under slow freezing conditions at −25 °C, while small ice crystals (1–2 µm) form during rapid freezing in liquid nitrogen. The rapid freezing of cellular starchy food resulted in a better quality product than slow freezing immediately after freezing. Rapid freezing of non-cellular starchy food, however, produced a product that was only slightly better in quality than the slowly frozen product. After storage, the rapidly frozen cellular product was still better in quality than the slowly frozen product, although the difference was smaller. The slight advantage gained by rapid freezing of the non-cellular product was lost during storage.

Undue emphasis on the importance of freezing speed is sometimes misleading. Unless freezing is excessively slow, days or weeks rather than hours, most products are comparatively insensitive to the speed of freezing. In any case, an increase in volume of around 10% is associated with freezing of most foods. Broadly speaking, faster is marginally better than slower in most of the food products. This is particularly true for fruit and vegetable products, but less so for animal tissue. Moreover, the initial advantage obtained by fast freezing is lost during storage due to recrystallization. Although fast freezing has advantages, some products will crack or even shatter if exposed directly to extremely low temperature for a long period of time. Researchers reviewed the fundamental aspects of freeze cracking and strategies for its prevention. The mechanisms to explain freeze cracking vary. The proposed mechanisms are:

- Volume expansion: The volume expansion due to the formation of ice and the amount of empty space in a microstructure are the primary factors affecting the degree of mechanical damage to cells during freezing. In addition, differences in moisture content, composition, or amount of un-freezable water may cause different degrees of cracks.
- Contraction and expansion: Cracking may also occur to relieve the product of internal stress from non-uniform contraction during rapid cooling. However, both contraction and expansion may cause freeze cracking.
- Internal stress: Fast freezing will cause crust formation at the surface, which serves as a shell and prevents further volume expansion when the internal portion of the unfrozen material undergoes freezing. This process then contributes to the internal stress build-up later in the freezing process. The freeze cracking will occur if the internal stress exceeds the strength of the exterior frozen material during processing. The distribution of the stress is the controlling factor, and it is governed by absorbing (dissipating) the stress into the structure or reflecting the stress to cause build-up of internal stress.

Generally, the circumferential tension in the outer surface of the muscle reached a breaking point, and a shallow crack formed along the length of the muscle or the surface yielded, causing a bulge. Researchers found that size, moisture content, density, modulus of elasticity, Poisson's ratio, and porosity all had significant influence on freeze cracking. However, no single property completely explained the development of freeze cracking. Excessive freezing speeds can ruin a food product.

The build-up of internal pressure during very rapid freezing shatters the already frozen external layers and produces very small crystals, leading to scattering of incident light. The current practice of quick freezing is generally chosen to save processing time (cost) and factory space. Moreover, rapidly chilled muscles become tough on freezing and thawing, a phenomenon known as cold shortening, which is not a problem in poultry.

In foods containing microbial cultures, it is important to maintain their activity. Rapid freezing causes detrimental effects on yeast activity of frozen dough. This may be due to the formation of intracellular ice crystals invariably lethal to yeast cell membranes. Yeast activity decreased significantly when the rate of cooling was increased from 0.98 to 1.57 °C/min.

Microbial Aspects

The detrimental effects of freezing on microorganisms may be desirable or undesirable, depending on the types of food products. In frozen foods without any added beneficial cultures, microbial growth or spoilage is not desirable, whereas care must be taken to reduce the damage in cells during freezing of foods containing microbial cultures. The maximum recommended storage temperature at which microbiological spoilage ceases is registered between -9 °C and -12 °C. Although microbiological spoilage can be avoided at these temperatures, the enzymes present in the product will still play a part in spoilage. Hence, hygienic conditions or heat processing (blanched or cooked) will increase the shelf life. Freezing causes the apparent death of 10%–60% of the viable microbe population, a percentage that gradually increases during frozen storage. Microorganisms differ considerably in their sensitivity to freezing; thus, the main concern is organisms that are likely to survive the freezing treatment and grow when the product is thawed. There is considerable variation in the ability of bacteria to resist damage by freezing. In general, Gram-negative bacteria are less resistant to freezing death than are Grampositive bacteria. Non-sporulating rods and spherical bacteria are the most resistant, while bacterial spores, such as Clostridium and Bacillus, remain unaffected by freezing. Bacteria in the stationary phase are more resistant than those in the log phase. Genera commonly encountered in frozen food include Pseudomonas, Achromobacter, Flavobacter, Micrococcus, Lactobacillus, Corynebacterium-like catalase-positive rods, enterococci, Streptococcus lactis, S. lactis-like streptococci, Aerococcus, and Pediococcus. Gianfranceschi and Aureli studied the survival of Listeria monocytogenes during freezing (-50 °C) and frozen storage (-18 °C) when inoculated into chicken breast, ground beef, spinach, mozzarella cheese, and codfish. They observed only a slight decrease in the viable population ranging from 0.1 to 1.6 log cycles after 57 min at -50 °C. Cells of L. monocytogenes were more resistant to death and injury when they were inoculated in ground beef and chicken breast, whereas they were less resistant in fish. A further reduction in viability of survival cells (up to 1.0 log) was detected after 240–300 days of storage at –18 °C.

The principal site of damage in the bacterial cells during freezing has been shown to be the membrane. Very rapid cooling of cells from room temperature to -150 °C resulted in more lipid crystallization before any rearrangement of intramembrane particles could occur. This leads to damage being mainly limited to the area around the cytoplasmic membrane. In slowly frozen samples, phase separation of the outer and cytoplasmic membranes was induced, causing the outer membrane to be split off by extracellular ice crystal formation. This damage could be reduced by the addition of a cryoprotectant, which modifies ice crystal formation. Damage to membranes leads to the leakage of internal cell materials, such as potassium ions, “-galactosidase, low molecular solutes, amino acids, RNA, and single- and double-strand DNA. The release of these substances has been correlated negatively with cell viability. Another type of damage is osmotic dehydration of cells caused by extracellular ice formation and the resulting increase in extracellular solute concentration. This process causes the intracellular macromolecules to move closer to the membranes. The development of repulsive forces gives rise to large anisotropic stresses in the membranes, resulting in deformation, phase separation, and formation of a non-lamellar phase. Moreover, salt addition and lowered pH also play a role in the complex nature of freeze injury and cell survival.

In fermented foods, such as yogurt, frozen storage should increase the viability of beneficial cultures incorporated for their potential health benefits and control of other spoiling microorganisms. The potential beneficial roles of bifidobacteria in the human intestine reported include antagonistic effects on enteropathogenic bacteria, breakdown of carcinogenic N-nitrosamines, and suppression of liver tumorigenesis. Thus, bifidobacteria are incorporated in dairy products. Kebary studied the viability of Bifidobacterium bifidum in the fermented dairy product Zabady. The number of bifidobacteria surviving after 5 weeks of frozen storage (-25 °C) of Zabady was higher ($>10^7$) than the minimum level necessary to achieve the beneficial effect of bifidobacteria ($10^5 - 10^6$ /mL). The total bacterial count decreased as the amount of added bifidobacteria increased. This could be due to the effect of antimicrobial substances produced by bifidobacteria. In eight strains of Lactobacillus acidophilus, higher rates and greater activity were always obtained by storing cultures at -80 °C, but most strains stored at -30 °C also survived well. The viability of frozen cultures was affected more by storage temperature than by cooling rate. The plate counts decreased less than 1 log cycle and fermentation activity was 40%–70% when cultures of Lactobacillus delbrueckii subsp. bulgaricus were stored at -80 °C for 1 year. However, fermentation activity was less than 10% when cultures were stored for 1 year at -30 °C. The fermentation activity of Streptococcus salivarius subsp. thermophilus was similarly reduced to 10%–60% after 1 year of storage at -30 °C. Protective solutes can be used to improve survival rates.

Yeast cells in bakery products do not withstand freezing well. This can be partially compensated for by increasing the amount of yeast used in the formulation or adding improved yeast strains having a better survival rate in freezing. The freeze-thaw tolerant yeast should have high trehalose content in addition to reduced activity. Trehalose has been reported to perform a cryoprotectant function in the yeast cell. Although the amount of yeast in the formulation could be increased, much higher levels of yeast (6%–8%) have a detrimental effect on aroma and flavor of the baked product. When dough pieces were made up and frozen immediately after mixing, yeast activity remained stable after prolonged storage periods. When fermentation was allowed to proceed after mixing and before freezing, the yeast became less tolerant to freezing temperature and its activity declined. This may be due to a change in the yeast cell membrane sensitivity.

Physical Changes and Quality

Free and Bound Water

Different types of water are present in frozen foods. These can be broadly categorized as free and un-freezable water, which does not freeze even at very low temperatures. A major cause of product degradation is the amount of unfrozen water present in the frozen matrix. Unfrozen water is known to be reactive, particularly during frozen storage, rendering the product susceptible to deteriorative and enzymatic reactions and limiting its frozen shelf life. The concept of glassy state is being applied to frozen foods stability, since molecular mobility reduces the reaction rates of the unfrozen water matrix and other components. Generally, un-freezable water molecules in aqueous solution are immobilized translationally or rotationally by solutes. The amount of un-freezable water can be measured experimentally and mathematically computed for different types of foods.

Weight Loss

Dehydration or weight loss should be regarded as an important quality parameter for frozen unpacked foods, mainly in animal tissue. Foods lose moisture during the freezing process because their surface is exposed to heat and a moisture gradient exists within the environment. Researchers monitored weight loss of meat during freezing and frozen storage and found a range of 0.28%–2.98% during the freezing process; meanwhile the global values (values including freezing and storage to reach up to 20 h of refrigeration) corresponded to a range between 1.67% and 6.15%. Thus, they demonstrated the importance of the surface dehydration on the meat quality.

Recrystallization

Ice recrystallization during frozen storage influences the product quality in different ways. Recrystallization of solutes and ice in frozen foods is also important to quality and shelf life. A polymer is least prone to crystallization at temperatures below glass. In general, the rate of recrystallization is highest at the midpoint between the melting and glass transition temperatures. Fluctuations in product temperature of 2 °C–3 °C, as are likely to be found in bulk cold stores kept at -18 °C or colder, are unlikely to cause perceptible damage even over long periods. However, frequent large fluctuations during retail display and during the carry-home period cause ice crystals to ripen or grow, coalesce, and move to the product surface. This leads ultimately to a freeze-dried product if packaging is permeable to moisture, allowing the sublimed or evaporated water vapor to escape. The loss of moisture results in toughening of animal tissue and greater exposure to any oxygen present.

Retrogradation

Quality loss by staling and starch retrogradation occurs most rapidly at chill temperatures in baked goods. After baking, starch from the loaf progressively crystallizes and loses moisture. Until a critical point of moisture loss is reached, freshness can be restored by heating and reabsorbing starch crystals. A tight wrap helps to keep the moisture content high for a certain amount of time. The complete crystallization of starch produces the crumbly texture of stale bread. Slow freezing is to be avoided to reduce the time spent at chill temperatures. Amylase is a useful antidote to bread staling. In general, moisture migration during frozen storage is the principal cause of staling. It has been reported that repeated freezing and thawing treatments favored starchy paste retrogradation, which is related to a mild hydrolysis of starch chains.

Protein Denaturation

Some protein denaturation and solubility changes are known to occur as a result of freezing, but the practical significance of these changes is not clear. Fish muscle has a unique arrangement of muscle fibers. It is divided into a number of segments called myotomes, which are separated from one another by a thin sheath of connective tissue called the myocomma or myoseptum. Fish deterioration during frozen storage is associated with a decrease of protein solubility and extractability, diminishing the nutritional value. Quality loss during cold storage of meat is characterized by an increase in loss of waterholding capacity, a decrease in protein extractability, a decrease in sulfhydryl groups, and a slight loss in ATPase activity. In frozen meat, water losses are related to the denaturation of myofibrillar protein. The water-holding capacity of the meat and the biochemical properties of actomyosin, such as enzymatic activity, viscosity, and surface hydrophobicity, are affected by freezing. The expressible moisture in adductor muscles increases during freezing and frozen storage. These changes are accompanied by actomyosin denaturation. The myosin and paramyosin of the actomyosin complex are most affected. An amino group from some essential amino acids, such as lysine, can react with the carbonyl group of reducing sugars during processing or storage. Peptides and amino acids are also increased in the drip fluid, as are nucleic acids, indicating protein changes and structural cellular damage. During freezing, water molecules freeze out and migrate to form ice crystals, resulting in the disruption of the organized H-bonding system that stabilizes the protein structure, and the hydrophobic as well as the hydrophilic regions of protein molecules become exposed to a new environment. This may allow the formation of intermolecular cross-linkages either within a protein molecule or between two adjacent molecules.

Freezer Burn

Moisture loss by evaporation from the surface of a product leads to freezer burn, an unsightly white color that can be mistaken for mold but is resolved on rehydration during cooking unless it is severe. It is usually in the form of patches of light-colored tissues, produced by evaporation of water, which leave air pockets between meat fibers. Dehydration of the product or freezer burn may occur while freezing an unpackaged food item in blast freezers unless the velocity of air is kept to about 2.5 m/s and the period of exposure to air is controlled. This dehydration can be controlled by humidification, lowering of storage temperatures, or better packaging. A single package of spinach and cauliflower experienced a 1.5-fold increase in loss of moisture per 2.8 °C rise in temperature between -17.8 °C and -6.7 °C. The loss of moisture occurs faster when held in a temperature cycle than when held at constant temperature during frozen storage.

Glass Formation

The glass transition has a dramatic effect on frozen food quality. The product is most stable below T_g', and moisture has little influence on T_g'. The presence of low molecular weight (LMW) solutes lowers the T_g', and high molecular weight (HMW) exerts little effect. This means that with increasing maturity, many vegetables display a decrease in sugars and an increase in starch, thus increasing the T_g' and the stability of frozen foods.

Functional Properties

Functional properties on any food product are normally affected by differences in freezing and thawing methods. Properties such as rheological (both flow and dynamic), textural, mechanical, consistency, appearance and sensory attributes and water-holding capacity have been correlated with freezing and thawing processes. The changes in specific functional properties become microscopically and qualitatively related to structural modifications or rearrangements of the food items. The rheological properties of fresh and frozen thawed okra dispersions were significantly different when measured at 20 °C–80 °C. The dispersions were pseudoplastic with both the consistency coefficient and the flow behavior index influenced by temperature. The consistency coefficient was higher for the unfrozen sample than the frozen thawed sample when measured at 20 °C and 50 °C, but the reverse was observed at 80 °C. The flow behavior indices were not different at any temperature between 20 °C and 80 °C. Researchers studied the effect of freezing rate on the rheological response of gelatinized starch pastes containing lipids. Low freezing rates increased the viscoelasticity, reduced the apparent viscosity, and led to higher structural changes of the pastes; in contrast, high freezing rates maintained the same rheological characteristics of the unfrozen samples. Researchers evaluated the viscoelastic behavior of refrigerated and frozen mozzarella cheese and found differences in the viscoelastic response. The storage modulus was higher for refrigerated samples in comparison with frozen cheese, in which modifications in cheese microstructure by freezing were observed with SEM methodology. Further, the freezing was correlated with casein hydrolysis, cheese matrix softening, and proteolysis acceleration.

Pickling

Pickling is one of the oldest methods of food preservation. The Chinese were fermenting vegetables as early as the third century BCE. By the first century CE, the Romans were pickling. Pickled products also appeared early in America, and the pickle barrel was common during the colonial days. Pickles even became part of our folklore, as children learned to recite the "Peter Piper picked a peck of pickled peppers" tongue twister. By the early 1920s, the U.S. Department of Agriculture (USDA) had published instructions on making pickles at home. Many of these procedures are still used today.

Preservation by Pickling

Microorganisms are always on vegetables. Proper home canning prevents the growth of the microorganisms that cause spoilage and illness. When the acidity of a canned food is high, harmful bacteria such as Clostridium botulinum cannot grow. That is why pickling (adding acid) prevents spoilage. There are two types of pickles:

- Brined (fermented) pickles require several weeks of "curing" at room temperature. During this period, colors and flavors change. Acid is produced as lactic acid bacteria grow.
- Quick (unfermented) pickles are made in 1 or 2 days by adding acid in the form of vinegar. It is critical to add enough vinegar to prevent bacterial growth.

Equipment for Fermenting

Crock

A 1-gallon container is needed for each 5 pounds of fresh vegetables. Therefore, a 5-gallon stoneware crock is an ideal size for fermenting about 25 pounds of fresh cabbage or cucumbers. Select a crock that is made for food use. Food-grade plastic and glass containers are excellent substitutes. Sauerkraut may be fermented in quart or half-gallon jars.

Caution: It is not safe to use garbage cans or other containers that are not made to come into contact with food for humans.

Weight

A weight keeps fermenting food under the brine. When using a crock, select a dinner plate or glass pie plate that is slightly smaller than the container opening. Weigh it down with 2 or 3 quart jars that are filled with water and have closed lids.

You also can use a large, food-grade plastic bag (either heavyweight or double-bagged) as a weight. Fill the bag with brine (1½ Tablespoons salt per 1 quart water) and close it. Place the bag in the fermentation container to cover and hold down the fermenting food. Using brine prevents dilution in case the bag leaks.

Cabbage and cucumbers must be kept 1 to 2 inches under the brine while fermenting. After weighing down, cover the container opening with a clean, heavy bath towel to prevent contamination from insects and molds during fermentation.

Other Equipment

- Utensils: Do not use zinc, copper, brass, galvanized metal, or iron utensils. These metals may react with acid or salt and affect the quality (color changes) and safety of pickles.
- Jars: Processing times are based on standard-sized, Mason-type jars designed for home canning use. Inspect jars for cracks and chips, and discard damaged ones.
- Lids: Two-piece, metal canning lids and screw bands are recommended. Follow the manufacturer's instructions for preparing the lids. To prevent sealing failures, do not reuse lids. Wash the screw bands and discard any with dents or rust.

- Boiling Water Canner: These canners, made of aluminum or porcelain-covered steel, have removable baskets and fitted lids. Any large container may be used if it has room for at least 1 inch of water to briskly boil over the tops of jars during processing. Place a metal rack in the bottom of the container to keep the jars from resting directly on the bottom, which can cause breakage.

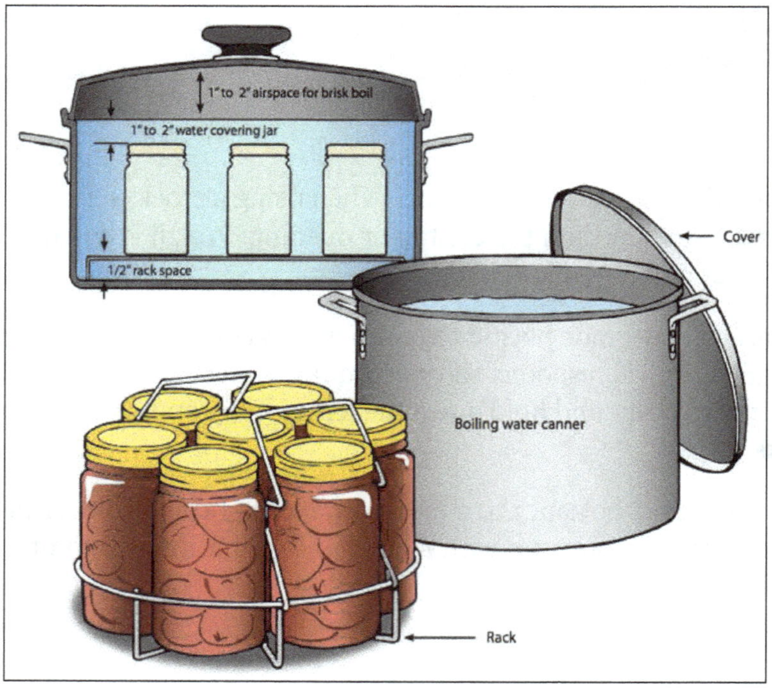

Boiling water canner.

Ingredients

Vegetables

Select tender vegetables without blemishes or mold.

Select tender vegetables without blemishes or mold and thoroughly wash them in cold water. When you select cucumbers, keep the following points in mind:

- Use varieties grown especially for pickling.
- Avoid commercially waxed cucumbers; pickling liquid cannot penetrate the wax easily.
- Pick cucumbers early in the morning and use within 24 hours, for best quality.
- Select cucumbers of similar size: 1½ inches long for gherkins,4 inches for dills. Use oddly shaped and more mature cucumbers for relishes and bread-and-butter style pickles.
- Trim off the blossom end of the cucumber; it may contain enzymes that soften the pickles.

Salt

- Pickling and canning salts are the best choices.
- Table salt may be used, but the additive that prevents caking may make the brine cloudy.
- Reduced-sodium salts (such as potassium chloride) may be used in quick pickle recipes; however, the pickles may have a slightly different taste.
- Do not use flake salt or kosher salt because they vary in density. Do not use rock salt or sea salt, which have impurities.

Caution: Do not use reduced-sodium salt in brined pickles or sauerkraut—these products need a specific amount of sodium to control bacterial growth and give a firm texture.

Vinegar

Vinegar is the most important ingredient in quick-pickle recipes. Without an adequate amount, the pickles will not be safe to eat. Use any vinegar with 5% acidity.

- White distilled vinegar has a pungent taste. Use it when pickling light-color foods such as cauliflower and onions.
- Cider vinegar has a milder flavor. It will make darker brine.

Caution: Do not reduce the amount of vinegar or increase the amount of water in a recipe. For a less sour taste, add a small amount of sugar. Do not use homemade vinegar. It varies in acidity.

Flavorings

Place flavorings at the bottom of a clean, suitable container.

- Spices should be fresh. Whole spices are best. Powdered spices may cause the pickles to darken and the brine to become cloudy.
- Dill may be used fresh, or as seeds or oil. The flavor of dill seeds may not be as good as fresh dill (1 teaspoon seeds equals 1 dill head). To use dill oil, substitute one drop for each head of dill. Fresh dill freezes well.

- Garlic should be fresh and at the peak of maturity. Immature or sprouting garlic can turn blue in the jar.
- Sugar may be either white or brown. Brown sugar may give a darker color to the brine. Corn syrup and honey may give an undesirable flavor.

Water

- Soft water makes the best brine for pickles.
- Hard water may cause cloudiness in the brine and discolor pickles. If only hard water is available, boil and let it sit for 24 hours. Skim off scum and use water from the top of the container without disturbing any sediment.

Firming Ingredients

If fresh, good quality ingredients are used and up-to-date procedures and processes are followed, you should not need firming agents to make crisp pickles. Soaking cucumbers in ice water for 4 to 5 hours prior to pickling is a safe method for making crisp pickles. If you choose to use firming agents, be sure to use food-grade products and follow the manufacturer's instructions.

- Pickling lime contains calcium. Cucumber or vegetable slices are sometimes soaked in a lime and water solution for 12 to 24 hours before pickling. Be sure to remove excess lime by rinsing and soaking the cucumbers for 1 hour in fresh, cold water and then repeating the process two more times. The excess lime must be removed.
- Calcium chloride for pickling is often sold where you buy canning supplies. It is usually added directly to the jars of pickles. Follow the instructions on the package for use.
- Alum may also be used to firm fermented cucumbers. However, it is not necessary and is not included in this publication. Alum does not improve the firmness of quick pickles.

Caution: Do not use lime sold at garden centers or lumberyards.

Packing the Jars

- Wash the jars in hot, soapy water and rinse well. Keep them hot until ready for use by placing them in the canner filled with hot water or in a hot dishwasher. Jars that will be filled with food and processed for less than 10 minutes should be sterilized by boiling them for 10 minutes. Follow manufacturer's directions for preparing lids.
- Prepare ingredients.
- Follow instructions for packing hot (vegetables preheated in brine) or raw ingredients into hot jars.
- Pack food and liquid to within 1/2 inch from the top of the jar, unless otherwise specified. If the jars are too full, some contents could siphon out of the jar during heat processing, stick to the rim, and prevent sealing. All air may not be exhausted during processing if the space between the packed food and liquid and the top of the jar (headspace) is too large.
- Run a plastic spatula around the inside of the jar to remove air bubbles. Adjust headspace, if needed.
- Wipe jar rims and threads with a clean, damp cloth to remove any bits of food that might prevent a seal.
- Adjust lids and screw bands.

Processing

Pickles are processed to stop fermentation and the growth of bacteria that cause spoilage. Use either conventional boiling water canner processing or lower-temperature pasteurization, if a firmer result is desired and the recipe allows for this method.

Caution: It is extremely important to wait until fermented pickles taste tart before processing.

Conventional, Boiling Water Canner Processing

- Arrange the filled jars on the rack of a canner filled halfway with very hot water (140 °F for raw-packed foods, 180 °F for hot-packed foods). Add more very hot water, if needed, to bring the water level up to 1 to 2 inches above jar tops. Place a tight-fitting cover on the canner.

- Turn heat to the highest setting and wait for the water to come to a full, rolling boil. When the water boils, set a timer for the recommended time. Increase processing times for higher altitudes, as directed, to make up for the lower boiling point of water. Keep water boiling gently and steadily. Add boiling water, if necessary, to keep jars covered.
- At the end of the processing time, turn off the heat and remove the canner lid. Wait 5 minutes before removing the jars. Spoilage could occur if jars are left in hot water.
- Using a jar lifter, remove the jars, being careful not to tilt them. Put jars on a rack or heavy towel so air can move freely around them. There should not be a cold draft or fan blowing on the jars. Let the jars sit undisturbed while they cool for 12 to 24 hours. Do not tighten the screw bands or press down on the center of the flat metal lid until the jars are completely cool.

Start timing when the water reaches a full, rolling boil.

Pasteurization at Lower Temperatures

- Processing at temperatures lower than the boiling point of water may result in a firmer or more crisp, final pickled product. Follow directions carefully to avoid possible spoilage. Use an instant-read thermometer or a candy, jelly, or dial thermometer that registers in the 120 °F to 185 °F range.
- Place filled jars on a rack in a canner half filled with warm water (120 °F to 140 °F). Add hot water to a level of 1 inch above jars.
- Heat the water to 180 °F to 185 °F and start a timer. Process for 30 minutes, checking with a candy or jelly thermometer to be certain that the water temperature stays at least 180 °F (temperatures higher than 185 °F may soften pickles).
- Immediately remove jars from the canner at the end of processing time.
- Using a jar lifter, remove the jars, being careful not to tilt them. Put jars on a rack or heavy towel so air can move freely around them. Let the jars sit undisturbed while they cool for 12 to 24 hours. Do not tighten the screw bands or press down on the center of the flat metal lid until the jars are completely cool.

Testing the Seals

Test each jar within 12 to 24 hours. Jars with flat, metal lids are sealed if:

- The lid has popped down in the center and does not move farther down when you press on it.
- The center of the lid gives a clear, ringing sound when you tap it with a spoon. Note: This is not as reliable as the first method.

Jars that have not sealed can be either refrigerated or reprocessed. To reprocess, start by removing the lid. Check and clean the sealing surface (the rim) of the jar. The brine in the jar must be reheated to boiling before reprocessing. Use a new lid and process again for the full time. The quality of the pickled food will not be as good after reprocessing.

Caution: If jars unseal during storage, do not consume.

Storing

Remove screw bands for storage so that the liquid under them will not cause rusting. Wipe the jars and label them with the date, contents, and processing information. Store jars in a cool, dark, and dry place.

After processing and cooling, store jars of pickles 4 to 5 weeks to develop ideal flavor. Use within 1 year for best eating quality and nutritive value. Exposure to heat, freezing, or light decreases the quality and shelf life of canned food.

Using Pickles

As you select each jar for use, examine it for signs of spoilage. Check the lid for tightness and vacuum seal. With the jar at eye level, examine for streaks of dried food on the outside of the jar. Inside of the jar, look for cloudy canning liquid, rising air bubbles, or any unnatural color. While opening the jar, watch for spurting liquid or cotton-like mold growth on food surfaces or the underside of the lid. Smell for unnatural or off odors.

Caution: Never taste food from a jar with an unsealed lid or food that shows signs of spoilage.

Carefully discard any jar of spoiled food to prevent possible illness to you, your family, or your pets. Any pickles not prepared according to these or other laboratory-tested recipes should be treated as having botulism and handled in one of two ways:

- If the suspect jars are still sealed, write on the jar:
    - Danger: Poison.
    - Do Not Eat.
    - Place the jars in a heavyweight garbage bag. Close and place the bag in a regular trash container or dispose of in a nearby landfill.
- If the suspect jars are unsealed, open, or leaking, they should be detoxified before disposal.

Detoxification Process

Wear rubber or heavy plastic gloves when handling suspect foods and cleaning up. Remove the jar lids and carefully place the suspect jars on their sides (along with the lids) in a stockpot, pan, or boiling water canner (8-quart volume or larger). Carefully add water to the pot until the level is 1 inch above the jars. Put a lid on the pot and heat the water to boiling. Boil for 30 minutes to detoxify the food. Cool and discard the food and lids. Wash jars and pot.

Spray or wet contaminated surfaces (such as counters and cutting boards) with a room-temperature household chlorine bleach solution (1 part unscented 5% to 6% sodium hypochlorite bleach to 5 parts clean water) and let stand for 30 minutes. Wipe up treated spills with paper towels and put them in a plastic bag before discarding in the trash. Rinse surfaces. Soak metal utensils in a mild chlorine solution (1 teaspoon bleach to 1 quart room-temperature water) for 30 minutes. Rinse. Wash your gloved hands well.

Food Sampling

To control food quality and acceptance within satisfactory limits, it is important to monitor the vital characteristics of raw materials, ingredients, and processed foods. This could be done by evaluating all foods or ingredients from a particular lot, which is feasible if the analytical technique is rapid and non-destructive. However, it is usually more practical to select a portion of the total product volume and assume the quality of the selected portion is typical of the whole lot.

Obtaining a portion, or sample, that is representative of the whole is referred to as sampling, and the total quantity from which a sample is obtained is called the population. Adequate sampling technique helps to ensure that sample quality measurements are an accurate and precise estimate of the quantity of the population. By sampling only a fraction of the population, a quality estimate can be obtained more quickly and with less expense and personnel time than if the total population were measured. The sample is only an estimate of the value of the population, but with proper sampling technique, it can be a very accurate estimate.

Sample Collection

It is important to clearly define the population that is to be sampled. The population may vary in size from a production lot, a day's production, to the contents of a warehouse. Extrapolating information obtained from a sample of a production lot to the population of the lot can be done accurately, but conclusions cannot be drawn from data describing larger populations, such as the whole warehouse.

Populations may be finite, such as the size of a lot, or infinite, such as in the number of temperature observations made of a lot over time. For finite populations, sampling provides an estimate of lot quality. In contrast, sampling from infinite populations provides information about a process. Figure represents the sampling for physical and chemical characteristics of food. Regardless of the population type, that is, finite or infinite, the data obtained from sampling are compared to a range of acceptable values to ensure the population sampled is within specifications.

Homogeneous Versus Heterogeneous Populations

The ideal population would be uniform throughout and identical at all locations. Such a population would be homogeneous. Sampling from such a population is simple, as a sample can be taken from any location and the analytical data obtained will be representative of the whole. However, this occurs rarely, as even in an apparently uniform product, such as sugar syrup, suspended particles and sediments in a few places may render the population heterogeneous. In fact, most populations that are sampled are heterogeneous. Therefore, the location within a population where a sample is taken will affect the subsequent data obtained. However, sampling plans and sample preparation can make the sample representative of the population or take heterogeneity into account in some other way.

Manual Versus Continuous Sampling

To obtain a manual sample the person taking the sample must attempt to take a "random sample" to avoid human bias in the sampling method. Thus, the sample must be taken from a number of locations within the population to ensure it is representative of the whole population. For liquids in small containers, this can be done by shaking prior to sampling. When sampling from a large volume of liquid, such as that stored in silos, aeration ensures a homogeneous unit. Liquids may be sampled by pipetting, pumping, or dipping. However, when sampling grain from a rail car, mixing is impossible and samples are obtained by probing from several points at random within the rail car. Such manual sampling of granular or powdered material is usually achieved with triers or probes that are inserted into the population at several locations. Errors may occur in sampling, as rounded particles may flow into the sampling compartments more easily than angular ones. Similarly, hygroscopic materials flow more readily into the sampling devices than does non-hygroscopic material. Horizontal core samples have been found to contain a larger proportion of smaller-sized particles than vertical ones. Continuous sampling is performed mechanically.

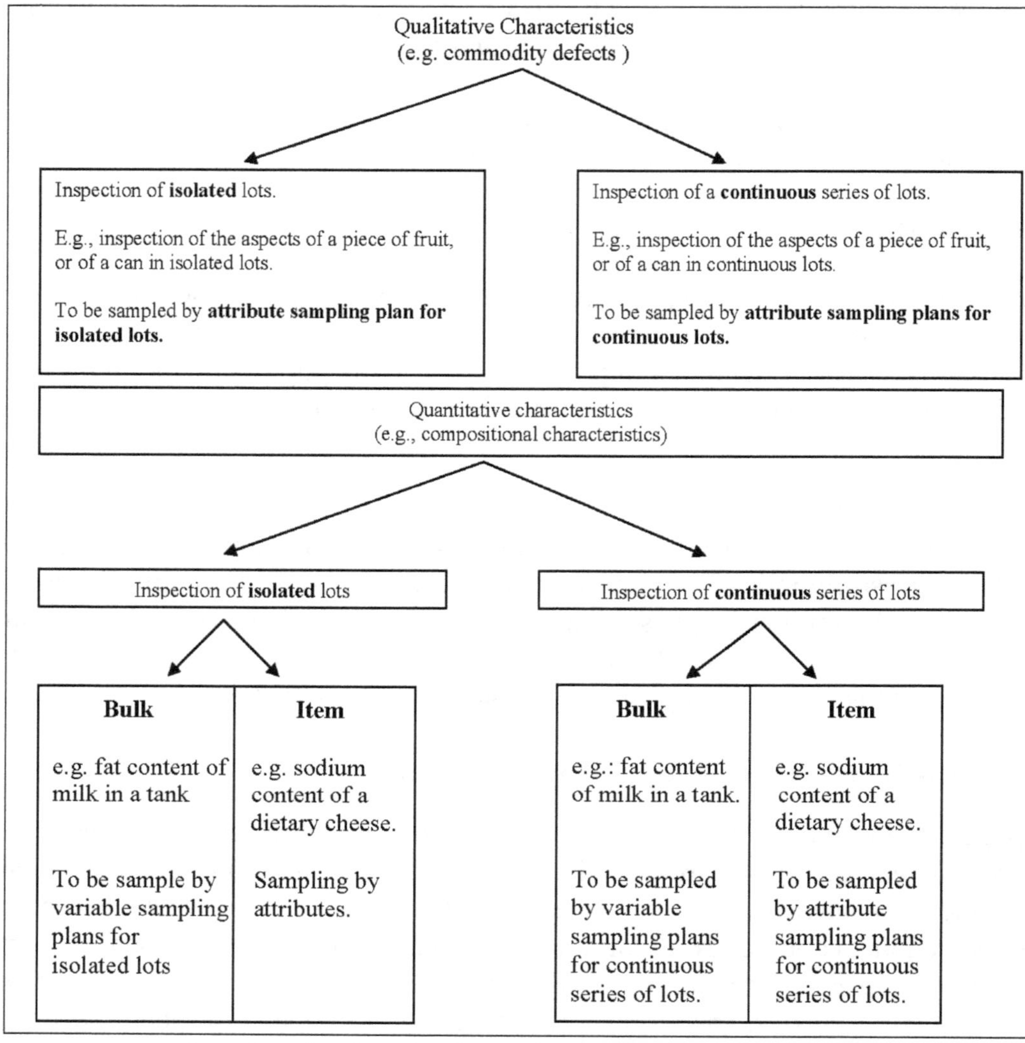

Sampling Flowchart for Chemical and Physical Characteristics.

Importance of Sample Collection

The reliability of analytical data thus obtained depends on several factors, sampling being the major factor. Current analytical methods require only few grams of food sample to analyze. Thus, it is necessary that a sample be as representative of the population as possible. There are three basic activities involved in analysis of food products:

- Collection of representative sample.
- Sample preparation.
- Analysis using appropriate methods and instruments.

These activities, although independent in nature, yet can have decisive influence on each other. Furthermore, each of these activities have their own potential sources of variations that contribute to the uncertainty level associated with any analytical result. Thus, care must be taken to identify the sources of variation and minimize or avoid them while accomplishing any activity. On the part of the laboratories, it is therefore necessary to develop a plan for the proper performance of each activity, and then establish quality standards and written procedures in compliance with the standards. Many times, the activity of sampling falls outside the purview of a laboratory's mandate or control. This is especially true in commercial testing laboratories where the "first contact" is the arrival of samples. To improve the overall quality of the analytical process, a laboratory must do all it can to receive appropriate, applicable, defensible samples. The development of appropriate plans will depend upon an understanding of the problems involved in each activity, and then the application of reasonable judgements in seeking solutions.

It should be noted that sampling terminology and procedures used may vary between companies and between specific applications. However, the principles described in this Unit are intended to provide a basis for understanding, developing, and evaluating sampling plans and sample handling procedures for specific applications encountered.

proper sampling, the analysts need to be consulted time to time concerning proper sample size, suitable containers for sampling or the use of appropriate preservatives to prevent any spoilage or transformation in a sample before analysis. One common cause of lack of precision or lab-to-lab variation in analytical results for a particular population can be traced back to erroneous sampling. For example, in case of grapes, a laboratory sample size of meager 3 kg berries represents the whole population of > 10000 kg in 1 hectare vineyard area. Thus, if the sample collected is not representative, then there will be sample-to-sample variation in results. When significant difference in results occurs among laboratories which have supposedly analyzed the same sample, a serious conflict may arise questioning the competence and credibility of the laboratories. Many of these situations can be avoided if samples are collected according to a rational plan that gives some assurance that the sample delivered to the laboratory represents the composition of the parent lot.

There are at least two ways to measure a given lot of goods: one, that we often assume to be the "proper" way, is to find its "true value", by which we mean its average value. The other way, often discovered accidentally as a result of "poor" sampling, is to measure its variability. So called proper sampling of drug dosage forms, for example, may involve compositing 20 tablets, by which the majority of the tablets could be used to dilute and conceal the fact that several of them are severely sub- or super- potent. Similarly, two lots of grain may have been purposely, but ineffectively, mixed in an attempt to reduce the average level of a contaminant. Sampling that led to the laboratory finding inconsistent results would reveal the attempt to dilute an illegal product.

Errors in Analytical Results due to Improper Sampling

Few studies have been conducted on the distribution of error among the three activities: sampling, sample preparation, and analysis. In one such study, which involved analysis of 20 nanogram/gram concentration of aflatoxin in a lot of peanuts, the error contributed by the sampling step was as high as 67% of the total variance, in comparison to 20% and 13% errors contributed by the analyst and the analytical procedure, respectively. In a field study conducted at the National Research Centre for Grapes, Pune, the pesticide residues in grape samples analyzed in 15 individual grape bunches collected out of 1 acre area showed above 50% sampling-induced variations. The results of such experimentations are not unusual and it illustrates the proportion of error that can be attributed to sampling. For peanuts, the distribution of aflatoxin can vary widely, with a few peanuts accounting for most of the contamination. Similarly, in case of the field sampling of grapes, the pesticide residues might have deposited in variable concentrations in different grape bunches and thus when they were analyzed separately, showed variable results. The important point in these examples is to show that sampling error can play a very significant part in the overall error in the analytical system.

Risks Associated with Sampling

There are two types of risks associated with sampling. Both should be considered when developing a sampling plan. The consumer risk describes the probability of accepting a poor quality population. This should happen rarely (<5% of the lots) but the actual acceptable probability of a consumer risk depends on the consequences associated with accepting an unacceptable lot. These may vary from major health hazards and subsequent fatalities to a lot being of slightly lower quality than standard lots. Obviously, the former demands a low or no probability of occurring whereas the latter would be allowed to occur more frequently. The second risk i.e., vendor risk is the probability of rejecting an acceptable product. As with consumer risk, the consequences of an error determine the acceptable probability of the risk. An acceptable probability of vendor risk is usually 5-10%.

Sampling Standards

Data obtained from an analytical technique are the result of a stepwise procedure from sampling, to sample preparation, laboratory analysis, data processing, and data interpretation. There is a potential for error at each step and the uncertainty, or reliability, of the final result depends on the cumulative errors at each stage. Variance is an estimate of the uncertainty. The total variance of the whole testing procedure is equal to the sum of the variances associated with each step of the sampling procedure and represents the precision of the process. Precision is a measure of the reproducibility of the data. In contrast, accuracy is a measure of how close the data are to the true value. The most efficient way to improve accuracy is to improve the reliability of the step with the greatest variance. Frequently, this is the initial sampling step. The reliability of sampling is dependent more on the sample size than on the population size. The larger the sample size, the more reliable the sampling. However, sample size is limited by time, cost, sampling methods, and the logistics of sample handling, analysis, and data processing.

It should be noted that sampling terminology and procedures used may vary between companies and between specific applications. Several standards and recommendations provide the ways and means to sample a particular lot. ISO specifies various standards and guidelines for drawing samples and data interpretation for samples. Few such standards related to food are mentioned in Table.

Table: Standards and Guides of Sampling.

ID of Standard/Guide	Title of Standard/Guide
ISO 2854 : 1976	Statistical Interpretation of data- techniques of estimation and tests relating to means and variances.
ISO 2859-0 : 1995	Sampling procedures for inspection by attributes-Part 0 ; Introduction to the ISO 2859 Attribute sampling system.

ISO 2859-1 : 1999 / IS 2500 (Part I) : 1989	Sampling procedures for inspection by attributes-Part 1; Sampling plans indexed by Acceptable Quality Level (AQL) for lot-by-lot inspection.
IS 2500 (Part II) : 1965	Sampling inspection procedures.
ISO 2859-2 : 1985	Sampling procedures for inspection by attributes- Part 2; Sampling plans indexed by Limiting Quality (LQ) for isolated lot inspection.
ISO 3494 : 1976	Statistical interpretation of data – Power of tests relating to mans and variances.
ISO 3951 : 1989	Sampling procedures and charts for inspection by variables for per cent non-conforming.
ISO 5725-1 : 1994	Application of statistics- Accuracy (trueness and precision) of measurement methods and results- Part 1; General principles and definitions.
ISO 7002 : 1986	Agricultural food products-Layout for a standard method of sampling a lot.
ISO 8423 : 1991	Sequential sampling plans for inspection by variables for per cent non-conforming (known standard deviation).
ISO 8422 : 1991	Sequential sampling plans for inspection by attributes.
ISO/ TR 8550 : 1994	Guide for the selection of an acceptance sampling system, scheme or plan for inspection of discrete items in lots.
ISO 10725 : 2000	Acceptance sampling plans and procedures for the inspection of bulk material.
ISO/ FDIS 11648/1	Statistical aspects of sampling from bulk materials – Part 1; General principles.
ISO/ DIS 14560	Acceptance sampling procedures by attributes – specified quality levels in non-conforming items per million.
Other IS specifications	Specifications published by BIS pertaining to individual products contain the details of lot size and corresponding number and quantity of samples to be drawn and tested for conformance.

Since all standards are subject to revision, parties to agreements based up on these guidelines should ensure that the most recent editions of the standards are always applied.

An example for flour sampling as per AOAC method is presented here. The AOAC Method 925.08 describes the method for sampling flour from sacks. The number of sacks to be sampled is determined by the square root of the number of sacks in the lot. The sacks to be sampled are chosen according to their exposure. The samples that are more frequently exposed are sampled more often than samples that are exposed less. Sampling is done by drawing a core from a corner at the top of the sack diagonally to the center. The sampling instrument is a cylindrical, polished trier with a pointed end. It is 13 mm in diameter with a slit at least one third of the circumference of the trier. A second sample is taken from the opposite corner in a similar manner. The cores are stored for analysis in a clean, dry, airtight container that has been opened near the lot to be sampled. The container should be sealed immediately after the sample is added. A separate container is used for each sack. Additional details regarding the container and the procedure also are described below.

Title 21 CFR specifies the sampling procedures required to ensure that specific foods conform to the standard of identity. In the case of canned fruits, 21 CFR 145.3 defines a sample unit as "container, a portion of the contents of the container, or a composite mixture of product from small containers that is sufficient for the testing of a single unit". Furthermore, a sampling plan is specified for containers of specific net weights. The container size is determined by the size of the lot. A specific number of containers must be filled for sampling of each lot size. The lot is rejected if the number of defective units exceeds the acceptable limit. For example, out of a lot containing 48,001 to 84,000 units, each weighing 1 kg or less, 48 samples should be selected. If six or more of these units fail to conform to the attribute of interest the lot will be rejected. Based on statistical confidence intervals, this sampling plan will reject 95% of the defective lots examined, that is, 5% consumer risk.

Table: Minimum Number of Primary Samples to be taken from a Lot.

Name of Commodity	Minimum Number of Primary Samples to be taken from a Lot
1. Meat and Poultry	
a) Non-suspect lot	1
b) Suspect lot	Determined according to Table.
2. Other products	
a) Products packaged in bulk, which can be assumed to be well mixed or homogeneous.	1 (A lot may be mixed by grading or manufacturingprocess)
b) Products packaged in bulk, which may not be well mixed or homogeneous.	For products comprised of large units, the minimum number of primary samples should comply with the minimum number of units required for the laboratory sample.
Either Weight of lot, kg	
<50	3
50-500	5
>500	10
or, Number of cans, cartons, containers in the lot	
1-25	1
26-100	5
> 100	10

Table: Number of randomly selected primary samples required for a given probability of finding at least one non-compliant sample in a lot.

Incidence of Non-compliant Residues in the Lot	Minimum number of samples (no) required to detect a non- compliant residue with a probability of		
%	90%	95%	99%
90	1	-	2
80	-	2	3
70	2	3	4
60	3	4	5
50	4	5	7
40	5	6	9
35	6	7	11
30	7	9	13
25	9	11	17
20	11	14	21
15	15	19	29
10	22	29	44
5	45	59	90

1	231	299	459
0.5	460	598	919
0.1	2302	2995	4603

- The table assumes random sampling.
- Where the number of primary samples indicated in Table is more than about 10% of units in the total lot, the number of primary samples taken may be fewer and should be calculated as follows:

$$n = \frac{n_0}{1 = (n_0 - 1) / N}$$

Where, n = minimum number of primary samples to be taken.

n_0 = number of primary samples given in Table.

N = number of units, capable of yielding a primary sample in the lot.

- Where a single primary sample is taken, the probability of detecting a non-compliance is similar to the incidence of non-compliant residues.
- For exact or alternative probabilities, or for a different incidence of non-compliance, the number of samples to be taken may be calculated from:

$$1 - p = (1 - i)^n$$

Where, p is the probability and i is the incidence of non-compliant residue in the lot (both expressed as functions, not percentages) and n is the number of samples.

Table: Sampling Procedure and Minimum Amount (composite/bulk) to be Sampled from lots.

Crop Type	Sampling Procedure	Example	Minimum Quantity
Root, tuber and bulb vegetables	Take samples from all areas of the crop. Remove as much adhering soil as possible from samples but do not wash.	Beet (red, sugar, fodder), onions, parsnips, potatoes, sweet potatoes, turnips	5 kg (and not less than 5 items)
	(Note: In some cases, where leaf parts are used as stock feed, they may need to be sampled separately).	Carrots, radish, spring onions.	2 kg
	Take the sample from all areas of the crop. Sample parts of the crop exposed to the spray and also those apparently protected by foliage.	Brassica (cabbage, cauliflower, broccoli, kohlrabi, curly kale).	5 kg (and not less than 5 items)
Leafy, stem, fruiting and legume vegetables	Remove as much soil as possible from crops such as celery, but do not wash.	Asparagus, brussel sprouts, celery, chicory, lettuce, spinach, turnip tops	2 kg
		Cucumber, melon, squashes, eggplant	5 kg (and not less than 5 items)
		Peppers, tomatoes, gherkins	2 kg
		Beans, peas, etc. (with pods)	2 kg

All tree and bush fruit, Including vines, small fruits and berries	Select fruit from all parts of the tree/ bush, high and low, and from both sides of the row, and select fruits according to abundance whether in each segment or the whole tree/bush. More fruit should therefore be selected from the more densely laden parts of the crop.	Apples, citrus, eaches, Pears	5 kg
	Sample parts of the crop exposed to the spray and also those apparently protected by foliage.	Cherries, nuts, olives, plums.	2 kg
	Take large and small fruits, perfect or slightly blemished, but not so small or blemished that they would not normally be saleable.	Bush fruit (all types), grapes, strawberries	2 kg
Cereal Grains	Cut not less than ten small areas (approximately 0.1 m^2) chosen randomly from all areas of the crop. Cut stalks about 10 cm above the ground. Remove grain from the straw.	Maize (grain and cobs)	2 kg
Oil Seeds	Collect the heads when they have reached the stage of maturity at which they are normally harvested and if convenient thresh to remove the seeds.	Sesame, canola, soybeans, sunflower	1 kg of seeds

Table: Description of Primary samples and min. size of laboratory sample.

S. No	Commodity Classification	Examples	Nature of Primary Samples to be taken	Minimum Size of Lab Sample
1. Primary food commodities of animal origin				
1.1	Large mammals Whole or half carcass Usually 10 kg or more	Cattle Sheep Pigs	Whole or part of diaphragm, supplemented by cervical muscles, if Necessary	0.5 kg
1.2	Small mammals Whole carcass	Rabbits	Whole carcass or hind quarters	0.5 kg, after removal of skin and bone
1.3	Mammal meat parts, loose fresh/chilled/frozen Packaged or other wise	Quarters Chops Steaks Shoulders	Whole unit(s),or a portion of a large unit	0.5 kg, after removal of bone
1.4	Mammal meat parts, bulk frozen	Quarters Chops	Either a frozen cross- section of a container or the whole (or portions of individual) meat parts	0.5 kg, after removal of bone
2. Poultry fats				
2.1	Birds, at slaughter whole or part-carcass	Chickens Turkeys	Units abdominal fat from at least three birds	0.5 kg
2.2	Bird meat parts	Legs Breast – muscle	Either visible fat, trimmed from units	0.5 kg
			Whole units or portions where fat is not trimmable	2 kg

2.3	Bird fat tissue in bulk		Units taken with a sampling device from at least three portions	0.5 kg
2.4	Poultry eggs		Whole eggs	12 whole eggs
2.5	Liquid, frozen or dried egg products		Whole units	0.5 kg
3. Processed foods of animal origin				
3.1	Mammal or bird comminuted, cooked, canned, dried, rendered or otherwise processed products including multi ingredient products	Ham sausage / Minced beef / Chicken paste	Packaged units or a representative cross section from a container or units (including juices, if any) taken with a sampling device	0.5 kg or 2 kg if fat content < 5%
3.2	Liquid milks, milk powders, ice creams etc.		Packaged units	0.5 L or 0.5 kg
3.3	Cheese Units < 0.3 kg	Units > 0.3 kg	Whole unit	0.5 kg
			Whole unit	0.3 kg
4. Herbs		Fresh	Whole units	0.5 kg
		Dried	Whole units	0.1 kg
5. Feed commodities of plant origin				
5.1	Legume animal feeds and other forages and fodders		Whole units	1 kg (at least 10 units)
5.2	Straw, hay and other dry products		Units taken with a sampling device	0.5 kg (at least 10 units)
6. Processed foods of plant origin				
6.1	Products of high unit value		Packages or units taken with a sampling device	0.1 kg
6.2	Solid products of low bulk density	Hops Tea	Packages or units taken with a sampling device	0.2 kg
6.3	Other solid products	Bread Flour Dried fruits Vegetable	Packages or units taken with a sampling device	0.5 kg
6.4	Liquid products	Oils Juices	Packages or units taken with a sampling device	0.5 L or 0.5 kg

The Sampling Plan

Understanding a Sample Plan

Sampling is generally done for a specific purpose and the purpose may indeed suggest or dictate the nature of any sampling plan. The International Union of Pure and Applied Chemistry (IUPAC) defines a sampling plan as "a predetermined procedure for the selection, withdrawal, preservation, transportation, and preparation of the portions to be removed from a lot as samples". A sampling plan should be a well-organized document that establishes the required procedures for accomplishing the program's objectives. It should address the issues of who, what, where, why, and how. The primary aim of sampling is to obtain a sample, subject to constraints on size that will satisfy the sampling plan specifications. A sampling plan should be selected on the basis of the sampling objective, the study population, the statistical unit, the sample selection criteria, and the analysis procedures. Factors determing the choice of a sampling plan are enlisted in Table. The two primary objectives of sampling are often to estimate the average value of a characteristic and determine if the average value meets the specifications defined in the sampling plan. The presence of a well-designed plan is important because it provides a consistent model to guide people performing the sampling activity, and it serves as a reminder of the important elements in this part of the overall sample analysis program.

Table: Factors Affecting Choice of Sampling Plans.

Factors to be considered		Questions
Purpose of inspection	a	Is it to accept or reject the lot?
	b	Is it to measure the average quality of the lot?
	c	Is it to determine the variability of the product?
Nature of Product	a	Is it homogeneous or heterogeneous?
	b	What is the unit size?
	c	How consistently have past populations met specifications?
	d	What is the cost of the material being sampled?
Nature of the test method	a	Is the test critical or minor?
	b	Will someone become sick or die if the population fails to pass the test?
Nature of the population being investigated	c	Is the test destructive or non-destructive?
	d	How much does the test cost to complete?
	a	Is the lot large but uniform?
	b	Does the lot consist of smaller, easily identifiable sublots?
	c	What is the distribution of the units within the population?

Statistical Approaches

In many sampling programs, statistical approaches are not given the requisite attention. Percentage sampling systems that specify a fixed percentage of a lot, say 5 or 10%, do not provide the quality protection that is often assumed. Statistical sampling theory furnishes the means to analyze the relationship between a lot of goods and the samples that are drawn from it. It can be used to estimate population measure, or "parameters," such as variance and correlation, from knowledge of corresponding samples quantities. The importance of sampling is recognized in ISO 17025, which requires that test reports make reference to the sampling procedure used by the laboratory or the submitting body.

Sampling Techniques/Methods

There are several sampling methods/techniques in common use. These are probability sampling, non-probability sampling, bulk sampling, and acceptance sampling. These are described in brief below:

Probability Sampling

Probability sampling is used when a representative sample is desired, and uses principles of statistical sampling and probability i.e. elimination of human bias. It is a random selection approach that tends to give each unit an equal chance of being selected. Simple random sampling requires that the number of units in the population be known and each unit is assigned a number. A specific quantity of random numbers between one and total number of population units is selected. Sample size is determined by lot size and potential impact of a consumer or vendor error. Units corresponding to the random numbers are then analyzed as an estimate of the population. Systematic sampling is used when a complete list of sample units is not available, but when samples are distributed evenly over time or space, such as on a production line. The first sample is selected at random and then every nth unit after that.

Stratified sampling involves dividing the population into overlapping subgroups so that each subgroup is as homogenous as possible. Group means, therefore, differ from each other as much as possible. Random samples are then taken from each subgroup. The procedure provides a representative sample because no part of the population is excluded and it is less expensive than simple random sampling.

Cluster sampling entails dividing the population into clusters or subgroups so that cluster's characteristics are as identical as possible, that is, the means are very similar to each other. Any heterogeneity occurs within each cluster. Clusters should be small and having a similar number of units in each cluster. The clusters are sampled randomly and may be either totally inspected or sub-sampled for analysis. This sampling method is more efficient and less expensive than simple random sampling, if populations can be divided into homogenous groups.

Composite sampling is used to obtain samples from bagged products such as flour, seeds, and larger items in bulk. Two or more samples are combined to obtain one sample for analysis that reduces differences between samples. For example, FDA composite 12 and at least six subsamples, respectively, for the sample to be analyzed for compliance with nutrition labeling regulations.

Non-probability Sampling

Non-probability sampling is used when it is not possible to collect a representative sample, or a representative sample is not desired. For example, in case of adulteration such as rodent contamination, the objective of the sampling plan may be to highlight the adulteration rather than collect a representative sample of the population. The sample collector uses judgement rather than statistical considerations in the selection of the sample. The unusual or unexpected characteristics in a population could be selected to be identified. This type of sampling is done in many ways, but in each case the probability of including any specific portion of the population is not equal because the investigator selects the samples without estimating sampling error.

Judgement sampling is solely at the discretion of the sampler and therefore is highly dependent on the person taking the sample. This method is used when it is the only practical way of obtaining the sample. This method may present a better estimate of the population than random sampling if sampling is done by an experienced individual and limitations of extrapolations from the results are understood.

Convenience sampling is performed when ease of sampling is the key factor. The first pallet in a lot or the sample that is most accessible is selected. This type of sampling will not be representative of the population, and therefore is not recommended. Restricted sampling may be unavoidable when the entire population is not accessible. For example, if sample is to be taken from a loaded truck, but the sample is not a representative of the entire population. Quota sampling is the division of a lot into groups representing various categories, and samples are then taken from each group. This method is less expensive than random sampling but also is less reliable.

Types of Sampling

Bulk Sampling

Bulk sampling involves the selection of a sample from a lot of material that does not consist of discrete, identifiable or constant units. Sampling may be performed in static or dynamic situations. Bulk sampling poses special problems requiring certain decisions to be made: the number of increments to be taken, the size of the increments, from where in the pile or stream they should be drawn, the sampling device to be used, and how to reduce the increments taken to a reasonable size of sample for delivery to the laboratory.

Acceptance Sampling

Acceptance sampling differs from the previous types and involves the application of a predetermined plan to decide whether a lot of goods meet defined criteria for acceptance. The risks of accepting "bad" or rejecting "good" lots are stated in conjunction with one or more parameters, for example, quality indices of the plan. Statistical plans can be designed to regulate the probabilities of rejecting good lots or accepting bad lots. There are two broad categories of acceptance sampling: sampling by attributes and sampling by variables.

Sampling by Attributes

In sampling by attributes, the unit of product is classified as defective or non-defective, or the number of defects in a unit of product is counted with respect to a given requirement. Or, the sampling is performed to decide on the acceptability of a population based on whether the sample possesses a certain characteristic, for example, Clostridium botulinum contamination in canned goods. An example of net weight determination may serve to explain the differences between the two categories. In attribute sampling, each unit that weighs 1 pound or more is accepted, and each unit that weighs less than 1 pound is rejected. If the number of rejects exceeds a predetermined number, the lot is rejected. If the number of rejects is less than the predetermined number, the lot is accepted.

Sampling by Variables

In variable sampling, sampling is performed to estimate quantitatively the amount of a substance (e.g., salt) or a characteristic (e.g., color) on a continuous scale. The estimate obtained from the sample is compared with an acceptable value (i.e., previously determined) and the deviation measured. This type of sampling usually produces data that have a normal distribution such as in the per cent fill of a container and total solids of a food sample. In general, variable sampling requires smaller sample size than attribute sampling and each characteristic should be sampled for separately when possible.

Operating Characteristic (OC) Curves

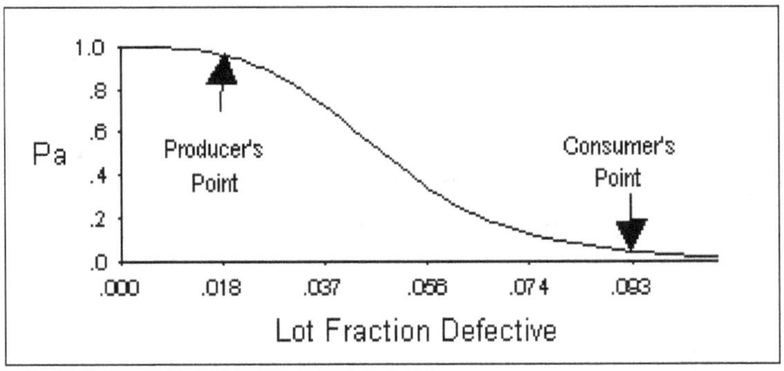

Operating Characteristic Curve

Operating Characteristic (OC) curves are used extensively in acceptance sampling. The OC curve shows the relationship between the quality and the per cent of lots expected to be acceptable for the quality characteristic inspected. In other words, the OC curve is a graph of lot defectives against the probability that the sampling plan will accept the lot. Figure depicts OC curves for an ideal sampling plan.

The Operating Characteristic (OC) curve shows the probability of acceptance, Pa, for any level of lot quality. On the horizontal axis is the quality characteristic. This OC curve enables you to evaluate the probability of acceptance for any true lot quality level-on a what-if basis. This way, you can design sampling plans that perform the way you want.

We can interpret the curve according to this example:

- If the lot quality is 0.093 fraction defective, then the probability of acceptance, Pa, is 0.05.
- If the lot quality is 0.018 fraction defective, then the probability of acceptance, Pa, is 0.95.

Requirements of Good Sampling Methods

Samples are useful for their intended purpose when they are taken in a manner consistent with generally recognized good sampling techniques and good sampling practices. This requires the following:

- Inspection of the lot before sampling.
- Use of suitable sampling devices for the particular commodity and type of sample desired.
- Use of suitable containers to hold the sample.
- Maintenance of the integrity of the sample and associated records.
- Use of adequate precautions in preserving, packing and delivery of the sample to the lab in a timely manner.
- Provision of appropriate storage conditions for the sample both prior to and following analysis.

All of these factors, along with others such as cost versus benefits analysis, and a review of program objectives and regularity requirements, are to be assessed and brought together in a sampling plan that serves as a guide to management, as well as to operating personnel as a firm plan to achieve quality in sampling.

Cost of Sampling

The attention of users is drawn upon relation between the efficiency and size of sample. For a given Acceptable Quality Level (AQL), the smaller the sample size, the smaller the cost of sampling, but the worse the efficiency, that is the risk to wrongly accepting a lot increases and worsens the damage in trade.

Problems in Sampling

Analytical data never are more reliable than the sampling technique. Sampling bias, due to non-statistically viable convenience, may compromise reliability. Errors also may be introduced by not understanding the population distribution and subsequent selection of an inappropriate sampling plan.

Unreliable data also can be obtained by non-statistical factors such as poor sample storage resulting in sample degradation. Samples should be stored in a container that protects the sample from moisture and other environmental factors that may affect the sample (e.g., heat, light, air). To protect against changes in moisture content, samples should be stored in an airtight container. Light sensitive samples should be stored in containers made of opaque glass, or the container wrapped in aluminum foil. Oxygen sensitive samples should be stored under nitrogen or an inert gas. Refrigeration or freezing may be necessary to protect chemically unstable samples. However, freezing should be avoided when storing unstable emulsions. Preservatives (e.g., mercuric chloride, potassium dichromate, and chloroform) can be used to stabilize certain food substances during storage.

Mislabeling of samples causes mistaken sample identification. Samples should be clearly identified by markings on the sample container in a manner such that markings will not be removed or damaged during storage and transport. For example, plastic bags that are to be stored in ice water should be marked with water-insoluble ink. If the sample is an official or legal sample the container must be sealed to protect against tampering and the seal mark easily identified. Official samples also must include the date of sampling with the name and signature of the sampling agent. The chain of custody of such samples must be identified clearly.

Three Class Sampling Plan

Another type of sampling used frequently by regulatory agencies to determine acceptance or rejection of a lot (often defined as the quality of product produced under essentially the same conditions but representing no more than one day's production) is the three-class sampling plan. This approach is often used when assessing microbiological contamination of foods. In this case, "n" is the number of samples, usually selected at random from the lot, the numerical value "m" represents acceptable concentrations, the numerical value "M" represents un-acceptable concentrations, and "c" is the maximum allowable number of marginally acceptable sample units such that if this number is exceeded, the lot is considered as un-acceptable. While "m" separates sample units of acceptable quality from those of marginally acceptable quality, "M" separates sample units of marginally acceptable quality from those of defective quality. For enforcement purposes, the sampling technique used should be the same as the sampling technique used to set the standard. For example, minimum reportable limits for particles are based on composite samples and not on individual lots.

Preparation of Sampling Plans

The development of quality sampling plans is a science in itself and has been given consideration by a number of organizations. One plan format that deserves serious consideration, developed by the International Organization for Standardization, is shown with comments in ISO/TC 34, ISO/DIS 7002.2, "Agricultural food products- Layout for a standard method of sampling from a lot ". It can serve as a starting point or check list for developing a sampling plan for most commodities. The title and headings from sections in the monograph are as below:

Model Sampling Plan

Agricultural food products- Layout for a standard method of sampling from a lot.

- Title (short but appropriate for index identification).
- Introduction (describing the purpose of the plan).
- Scope (describing the breadth of coverage of the plan).
- Field of application (products to be covered; where sampling will be done).
- References (documents, the validity of the plan with reference to other requirements).
- Definitions (specific terms associate with a particular matrix).
- Principle (statistical basis of the method of sampling).
- Administrative arrangements:
    - Sampling personnel.
    - Representation of parties concerned.
    - Health, safety and security precautions.
    - Preparation of the sampling report.
- Identification and inspection of the lot prior to sampling (important in survey sampling for identification, condition of the lot and selection of method of sampling).
- Sampling equipment and ambient conditions (proper tools such as use of sterile equipment for aseptic sampling).
- Sample containers and packing (essential to prevent contamination and damage during shipment or storage).
- Sampling procedures (as dictated by the plan objectives).
    - Sampling size.

    - Taking the sample.
    - Preparation of bulk samples and reduced samples.
    - Selection of samples of pre-packaged products.
- Packing, sealing and marking of samples and sample containers (identification of units and to establish chain-of-custody).
    - Filling and sealing sample containers.
    - Labeling or marking (including signature of sampling personnel).
    - Packing samples for storage or transportation.
- Precautions during storage and transportation of samples.
- Sampling report:
    - Administrative details.
    - Details of unit packs or enclosure containing the lot.
    - Material samples.
    - Marking and sealing of samples.
- Annexure (supplemental information, if necessary).

Sub Sampling for Analysis and Taking the Test Portion

If the test portion analyzed does not represent the sample or the lot from which it was taken, in that case, even the best analysis could give misleading information. Distortions introduced at this point will carry through the path of analysis and adversely affect the final results and the conclusions drawn from them. There are generally two choices in analytical sub sampling:

- Preparation of a composite laboratory sample (if multiple units are submitted for analysis).
- Examination of individual units.

Composite Lab Sample Preparation

A composite lab sample is one in which the individual units or representative portions of units are mixed to form a uniform mixture. Portions are then taken from the composite for analysis. Compositing saves analytical time and in some types of contract testing it may be the procedure specified. If the results indicate that there may be a problem, it will likely be necessary to go back and analyze individual samples. Compositing is not the procedure of choice when there is a chance that an individual unit that constitutes a public health or safety threats will not be detected (there are some exceptions) or where a unit at or outside of tolerance level will not be detected because of matrix dilutions. Multiple unit lab sampling is indicated when the possible range of values among individual units is considered significant or it is desirable to establish the variability of the lot.

Chapter 5

A Comprehensive Study of Food Storage

Food storage refers to the processes which help in storing food in such a manner that it can be utilized at a later stage for consumption. This chapter closely examines the key concepts of food storage to provide an extensive understanding of the subject. Some of the topics elaborated in this chapter are food bank, food storage containers, sweet potato storage, etc.

Food Storage

Yup'ik elevated food cache (qulvarvik), Hooper Bay, Alaska, 1929. Photograph by Edward S. Curtis.

Food logistics is both a traditional domestic skill and is important industrially. The transport and storing of food as well as timely delivery to consumers is important in order to secure the procurement of food. Food is stored by almost every human society and by many animals. Storing of food has several main purposes:

- Storage of harvested and processed plant and animal food products for distribution to consumers.
- Enabling a better balanced diet throughout the year.
- Reducing kitchen waste by preserving unused or uneaten food for later use.
- Preserving pantry food, such as spices or dry ingredients like rice and flour, for eventual use in cooking.
- Preparedness for catastrophes, emergencies and periods of food scarcity or famine.

- Religious reasons (Example: LDS Church leaders instruct church members to store food).
- Protection from animals or theft.

U.S. Federal Emergency Management Agency (FEMA) food storage containers stacked on shipping pallets in Texas, 2008.

History of Food Storage

Domestic Food Storage

Tupperware kitchen storage containers designed for a variety of uses.

The safe storage of food for home use should strictly adhere to guidelines set out by reliable sources, such as the United States Department of Agriculture. These guidelines have been thoroughly researched by scientists to determine the best methods for reducing the real threat of food poisoning from unsafe food storage. It is also important to maintain proper kitchen hygiene, to reduce risks of bacteria or virus growth and food poisoning. The common food poisoning illnesses include Listeriosis, Mycotoxicosis, Salmonellosis, E. coli, Staphylococcal food poisoning and Botulism. There are many other organisms that can also cause food poisoning.

Plastic storage containers can be used to store food.

There are also safety guidelines available for the correct methods of home canning of food. For example, there are specific boiling times that apply depending upon whether pressure canning or waterbath canning is being used in the process. These safety guidelines are intended to reduce the growth of mold and bacteria and the threat of potentially-fatal food poisoning.

Food Storage Safety

Freezers and Thawing Food

Freezer temperature should be maintained below 0 °F. Food should never be thawed at room temperature, this increases the risk of bacterial and fungal growth and accordingly the risk of food poisoning. Once thawed, food should be used and never refrozen. Frozen food should be thawed using the following methods:

- Microwave oven
- During cooking
- In cold water (place food in watertight, plastic bag; change water every 30 minutes)
- In the refrigerator

Throw out foods that have been warmer than 40 °F for more than 2 hours. If there is any doubt at all about the length of time the food has been defrosted at room temperature, it should be thrown out. Freezing does not destroy microbes present in food. Freezing at 0 °F does inactivate microbes (bacteria, yeasts and molds). However, once food has been thawed, these microbes can again become active. Microbes in thawed food can multiply to levels that can lead to foodborne illness. Thawed food should be handled according to the same guidelines as perishable fresh food.

Food frozen at 0 °F and below is preserved indefinitely. However, the quality of the food will deteriorate if it is frozen over a lengthy period. The United States Department of Agriculture, Food Safety and Inspection Service publishes a chart showing the suggested freezer storage time for common foods.

Refrigeration

It is important to note that safe food storage using refrigeration requires adhering to temperature guidelines:

For safety, it is important to verify the temperature of the refrigerator. Refrigerators should be set to maintain a temperature of 40 °F or below. Some refrigerators have built-in thermometers to measure their internal temperature. For those refrigerators without this feature, keep an appliance thermometer in the refrigerator to monitor the temperature. This can be critical in the event of a power outage. When the power goes back on, if the refrigerator is still 40 °F, the food is safe. Foods held at temperatures above 40 °F for more than 2 hours should not be consumed. Appliance thermometers are specifically designed to provide accuracy at cold temperatures. Be sure refrigerator/freezer doors are closed tightly at all times. Don't open refrigerator/freezer doors more often than necessary and close them as soon as possible.

Storage Times for Refrigerated Food

The United States Department of Agriculture, Food Safety and Inspection Service publishes recommended storage times for refrigerated food.

Storing Oils and Fats

Oils and fats can begin to go rancid quickly when not stored safely. Rancid cooking oils and fats do not often smell rancid until well after they have spoiled. Oxygen, light and heat all contribute to cooking oils becoming rancid. The higher the level of polyunsaturated fat that an oil contains, the faster it spoils. The percentage of polyunsaturated fat in some common cooking oils is: safflower (74%); sunflower (66%); corn (60%); soybean (37%); peanut (32%); canola (29%); olive (8%).

To help preserve oils from rancidification, they should be refrigerated once opened. Opened, refrigerated cooking oils should be used within a few weeks, when some types begin to go rancid. Unopened oils can have a storage life of up to one year, but some types have a shorter shelf-life even when unopened (such as sesame and flaxseed).

Dry Storage of Foods

Vegetables

A large root cellar at the Oxon Hill Manor farm in Maryland.

The guidelines vary for safe storage of vegetables under dry conditions (without refrigerating or freezing). This is because different vegetables have different characteristics, for example, tomatoes contain a lot of water, while root vegetables such as carrots and potatoes contain less. These factors, and many others, affect the amount of time that a vegetable can be kept in dry storage, as well as the temperature needed to preserve its usefulness. The following guideline shows the required dry storage conditions:

- Cool and dry: onion
- Cool and moist: root vegetable, potato, cabbage
- Warm and dry: winter squash, pumpkin, sweet potatoes, dried hot peppers

Many cultures have developed innovative ways of preserving vegetables so that they can be stored for several months between harvest seasons. Techniques include pickling, home canning, food dehydration, or storage in a root cellar.

Grain

Grain, which includes dry kitchen ingredients such as flour, rice, millet, couscous, cornmeal, and so on, can be stored in rigid sealed containers to prevent moisture contamination or insect or rodent infestation. For kitchen use, glass containers are the most traditional method. During the 20th century plastic containers were introduced for kitchen use. They are now sold in a vast variety of sizes and designs.

Metal cans are used (in the United States the smallest practical grain storage uses closed-top #10 metal cans). Storage in grain sacks is ineffective; mold and pests destroy a 25 kg cloth sack of grain in a year, even if stored off the ground in a dry area. On the ground or damp concrete, grain can spoil in as little as three days, and the grain might have to be dried before it can be milled. Food stored under unsuitable conditions should not be purchased or used because of risk of spoilage. To test whether grain is still good, it can be sprouted. If it sprouts, it is still good, but if not, it should not be eaten. It may take up to a week for grains to sprout. When in doubt about the safety of the food, throw it out as quickly as possible.

Spices and Herbs

Spices and herbs are today often sold prepackaged in a way that is convenient for pantry storage. The packaging has dual purposes of both storing and dispensing the spices or herbs. They are sold in small glass or plastic containers or resealable plastic packaging. When spices or herbs are homegrown or bought in bulk, they can be stored at home in glass or plastic containers. They can be stored for extended periods, in some cases for years. However, after 6 months to a year, spices and herbs will gradually lose their flavour as oils they contain will slowly evaporate during storage.

Spices and herbs can be preserved in vinegar for short periods of up to a month, creating a flavoured vinegar.

Alternative methods for preserving herbs include freezing in water or unsalted butter. Herbs can be chopped and added to water in an ice cube tray. After freezing, the ice cubes are emptied into a plastic freezer bag for storing in the freezer. Herbs also can be stirred into a bowl with unsalted butter, then spread on wax paper and rolled into a cylinder shape. The wax paper roll containing the butter and herbs is then stored in a freezer, and can be cut off in the desired amount for cooking. Using either of these techniques, the herbs should be used within a year.

Meat

Unpreserved meat has only a relatively short life in storage. Perishable meats should be refrigerated, frozen, dried promptly or cured. Storage of fresh meats is a complex discipline that affects the costs, storage life and eating quality of the meat, and the appropriate techniques vary with the kind of meat and the particular requirements. For example, dry ageing techniques are sometimes used to tenderize gourmet meats by hanging them in carefully controlled environments for up to 21 days, while game animals of various kinds may be hung after shooting. Details depend on personal tastes and local traditions. Modern techniques of preparing meat for storage vary with the type of meat and special requirements of tenderness, flavour, hygiene, and economy.

Semi-dried meats like salamis and country style hams are processed first with salt, smoke, sugar, acid, or other "cures" then hung in cool dry storage for extended periods, sometimes exceeding a year. Some of the materials added during the curing of meats serve to reduce the risks of food poisoning from anaerobic bacteria such as species of Clostridium that release botulinum toxin that can cause botulism. Typical ingredients of curing agents that inhibit anaerobic bacteria include nitrates. Such salts are dangerously poisonous in their own right and must be added in carefully controlled quantities and according to proper techniques. Their proper use has however saved many lives and much food spoilage.

Like the semi-dried meats, most salted, smoked, and simply-dried meats of different kinds that once were staples in particular regions, now are largely luxury snacks or garnishes; examples include jerky, biltong, and varieties of pemmican, but ham and bacon for instance, still are staples in many communities.

Food Rotation

Food rotation is important to preserve freshness. When food is rotated, the food that has been in storage the longest is used first. As food is used, new food is added to the pantry to replace it; the essential rationale is to use the oldest food as soon as possible so that nothing is in storage too long and becomes unsafe to eat. Labelling food with paper labels on the storage container, marking the date that the container is placed in storage, can make this practice simpler. The best way to rotate food storage is to prepare meals with stored food on a daily basis.

For Emergency Preparation

Guides for surviving emergency conditions in many parts of the world recommend maintaining a store of essential foods; typically water, cereals, oil, dried milk, and protein rich foods such as beans, lentils, tinned meat and fish. A food storage calculator can be used to help determine how much of these staple foods a person would need to store in order to sustain life for one full year. In addition to storing the basic food items many people choose to supplement their food storage with frozen or preserved garden-grown fruits and vegetables and freeze-dried or canned produce. An unvarying diet of staple foods prepared in the same manner can cause appetite exhaustion, leading to less caloric intake. Another benefit to having a basic supply of food storage in the home is for the potential cost savings. Costs of dry bulk foods (before preparation) are often considerably less than convenience and fresh foods purchased at local markets or supermarkets. There is a significant market in convenience foods for campers, such as dehydrated food products.

Commercial Food Logistics

Grain and beans are stored in tall grain elevators, almost always at a rail head near the point of production. The grain is shipped to a final user in hopper cars. In the former Soviet Union, where harvest was poorly controlled, grain was often irradiated at the point of production to suppress mold and insects. In the U.S., threshing and drying is performed in the field, and transport is nearly sterile and in large containers that effectively suppresses pest access, which eliminates the need for irradiation. At any given time, the U.S. usually has about two weeks worth of stored grains for the population.

Silos connected to a grain elevator on a farm in Israel.

Fresh fruits and vegetables are sometimes packed in plastic packages and cups for fresh premium markets, or placed in large plastic tubs for sauce and soup processors. Fruits and vegetables are usually refrigerated at the earliest possible moment, and even so have a shelf life of two weeks or less.

In the United States, livestock is usually transported live, slaughtered at a major distribution point, hung and transported for two days to a week in refrigerated rail cars, and then butchered and sold locally. Before refrigerated rail cars, meat had to be transported live, and this placed its cost so high that only farmers and the wealthy could afford it every day. In Europe much meat is transported live and slaughtered close to the point of sale. In much of Africa and Asia most meat is for local populations is raised, slaughtered and eaten locally, which is believed to be less stressful for the animals involved and minimizes meat storage needs. In Australia and New Zealand, where a large proportion of meat production is for export, meat enters the cold chain early, being stored in large freezer plants before being shipped overseas in freezer ships.

Food Bank

Volunteers pass out food items from a food bank run by Feeding America.

A food bank or foodbank is a non-profit, charitable organization that distributes food to those who have difficulty purchasing enough food to avoid hunger.

In North America and Australia, food banks usually operate on the "warehouse" model. They act as food storage and distribution depots for smaller front line agencies; and usually do not themselves give out food directly to the hungry. After the food is collected, sorted, and reviewed for quality, these food banks distribute it to non-profit community or government agencies, including *food pantries*, *food closets*, soup kitchens, homeless shelters, orphanages, and schools.

Outside North America and Australia, the "front line" model is often found. Such food banks give out most or all of their food directly to the end users. For both models, the largest sources of food include for-profit growers, manufacturers, distributors and retailers who in the normal course of business have excess food that they cannot sell. Some foodbanks receive a substantial proportion of their food from individual donors, including their volunteer workers. There is considerable overlap with food salvage, food rescue and gleaning.

The world's first food bank was established in the US in 1967, and since then many thousands have been set up all over the world. In Europe, which until recently had little need for food banks due to extensive welfare systems, their numbers grew rapidly after the lasting global inflation in the price of food which began in late 2006, and especially after the financial crisis of 2007–08 began to further worsen economic conditions for those on low income.

The growth of food banks has been broadly welcomed, most especially by those on the right of the political spectrum, but also by many on the left, who see them as evidence of active community that is independent of the state. However, academics and commentators have expressed concern that the rise of foodbanks may erode political support for welfare provision. Researchers have reported that food banks can be inefficient compared with state run services, and that some people feel ashamed at having to turn to them.

Standard Model

The warehouse of the Capital Area Food Bank.

With thousands of food banks operating around the world, there are many different models.

A major distinction between food banks is whether or not they operate on the "front line" model, giving out food directly to the hungry, or whether they operate with the "warehouse" model, supplying food to intermediaries like food pantries, soup kitchens and other front-line organisations. In the US, Australia and to an extent in Canada, the standard model is for food banks to act as warehouses rather than as suppliers to the end user, though there are exceptions. In other countries, food banks usually do hand out food parcels direct to hungry people, providing the service that in the US is offered by *food pantries*.

Another distinction is between the charity model and the labour union model. At least in Canada and the US, food banks run by charities often place relatively more weight on the salvaging of food that would otherwise go to waste, and on encouraging voluntarism, whereas those run by unions can place greater emphasis on feeding the hungry by any means available, on providing work for the unemployed, and on education, especially on explaining to users their civil rights.

In the US, cities will often have a single food bank which acts as a centralized warehouse and will serve several hundred front line agencies. Like a blood bank, that warehouse serves as a single collection and distribution point for food donations. A food bank operates a lot like a for-profit food distributor, but in this case it distributes food to charities, not to food retailers. There is often no charge to the charities, but some food banks do charge a small "shared maintenance" fee to help defray the cost of storage and distribution.

For many US food banks, most of their donated food comes from food left over from the normal processes of for-profit companies. It can come from any part of the food chain, e.g. from growers who have produced too much or whose food is not sufficiently visually appealing; from manufacturers who overproduced; or from retailers who over-ordered. Often the product is approaching or past its "sell by" date. In such cases, the food bank liaises with the food industry and with regulators to make sure the food is safe and legal to distribute and eat.

Volunteers weigh food drive donations.

Other sources of food include the general public, sometimes in the form of "food drives", and government programs that buy and distribute excess farm products mostly to help support higher commodity prices. Food banks can also buy food either at market prices or from wholesalers and retailers at discounted prices, often at cost. Sometimes farmers will allow food banks to send gleaners to salvage leftover crops for free once their primary harvest is complete. A few food banks have even taken over their own farms, though such initiatives have not always been successful.

Many food banks don't accept fresh produce, preferring canned or packaged food due to health and safety concerns, though some have tried to change this as part of a growing worldwide awareness of the importance of nutrition. As an example, in 2012, *London Food Bank* (Canada) started accepting perishable food, reporting that as well as the obvious health benefits, there were noticeable emotional benefits to recipients when they were given fresh food.

Summer can be a challenging time for food banks, especially in regions where school children are usually given regular free meals during term time. Spikes in demand can coincide with periods where donations fall due to folk being on holiday.

North America

History

In the U.S. and often also in Canada, food banks don't typically give food direct to the hungry. Instead they act as warehouses, supplying front line agencies like this Californian soup kitchen. (Pic taken in 2009, and shows members of the United States Navy serving visitors.)

The world's first food bank was the St. Mary's Food Bank Alliance in Arizona, founded by John van

Hengel in 1967. According to sociology professor Janet Poppendieck, hunger within the US was widely considered to be a solved problem until the mid-1960s. By the mid sixties, several states had ended the free distribution of federal food surpluses, instead providing an early form of food stamps which had the benefit of allowing recipients to choose food of their liking, rather than having to accept whatever happened to be in surplus at the time. However, there was a minimum charge and some people could not afford the stamps, leading to severe hunger. One response from American society to the rediscovery of hunger was to step up the support provided by soup kitchens and similar civil society food relief agencies - some of these dated back to the Great Depression and earlier. In 1965, while volunteering for a community dining room, van Hengel learned that grocery stores often had to throw away food that had damaged packaging or was near expiration. He started collecting that food for the dining room but soon had too much for that one program. He thought of creating a central location from which any agency can receive donations. Described as a classic case of "if you build it they will come", the first food bank was created with the help of St. Mary's Basilica. Food banks spread across the United States, and to Canada. By 1976, the precursor to Feeding America had been established. As of the early 21st century, their network of 200+ foodbanks provides support for 90,000 projects. Other large networks exist such as AmpleHarvest.org, created by CNN Hero Gary Oppenheimer which lists some 7,100 food pantries (1 out of every 5 in America) that can utilize overproduction of fresh produce.

It was not however until the 1980s that U.S. food banks began to enjoy rapid growth. A second response to the "rediscovery" of hunger in the mid sixties had been extensive lobbying of politicians to improve welfare. Until the 1980s, this approach had greater impact. In the 1970s, U.S. federal expenditure on hunger relief grew by about 500%, with food stamps distributed free of charge to those in greatest need. According to Poppendieck, welfare was widely considered preferable to grass roots efforts, as the latter could be unreliable, did not give recipients consumer-style choice in the same way as did food stamps, and risked recipients feeling humiliated by having to turn to charity. In the early 1980s, president Reagan's administration scaled back welfare provision, leading to a rapid rise in activity from grass roots hunger relief agencies. According to a comprehensive government survey completed in 2002, over 90% of food banks were established in the US after 1981. Poppendieck says that for the first few years after the change, there was vigorous opposition from left, who argued that state welfare was much more suitable for meeting recipients needs. But in the decades that followed, food banks have become an accepted part of America's response to hunger. Demand for the services of US food bank increased further in the late 1990s, after the "end of welfare as we know it" with President Clinton's Personal Responsibility and Work Opportunity Act. In Canada, foodbanks underwent a period of rapid growth after the cutbacks in welfare that took place in the mid-1990s. As early as the 1980s, food banks had also begun to spread from the United States to the rest of the world. The first European food bank was founded in France during 1984. In the 1990s and early 2000s, food banks were established in South America, Africa and Asia, in several cases with van Hengel acting as a consultant. In 2007, *The Global FoodBanking Network* was formed.

Food Aid for Pets

Some U.S. cities have organizations that provide dog and cat food for pets whose owners qualify for food assistance. For example, *Daffy's Pet Soup Kitchen* in Lawrenceville, Georgia is considered the largest pet food aid agency in Georgia, distributing over 800,000 pounds of dog and cat food in 2012. *Daffy's Pet Soup Kitchen* was started in 1997 by Tom Wargo, a repairman who was working in an elderly woman's home when he noticed her sharing her Meals On Wheels lunch with her pet cat because she couldn't afford cat food. Daffy's was one of seven non-profits recognized by Barefoot Wine in 2013 through a $10,000 donation and by being featured on labels of the vintner's Impression Red Blend wines. *Pet Buddies Food Pantry* in Woodstock, Georgia is another example of an establishment that provides food aid for pets.

Post 2007 Financial Crisis

Following the financial crisis of 2007–08, and the lasting inflation in the price of food that began in late 2006, there has been a further increase in the number of individuals requesting help from American and Canadian food banks. By 2012, according to *Food Banks Canada*, over 850,000 Canadians needed help from a food bank each month. For the United States, *Gleaners Indiana Food bank* reported in 2012 that there were now 50 million Americans struggling with food insecurity (about 1 in 6 of the population), with the number of individuals seeking help from Food banks having increased by 46% since 2005. According to a 2012 UCLA Center for Health Policy Research study, there has been a 40% increase in demand for Californian food banks since 2008, with even married couples who both work sometimes requiring the aid of food banks. Dave Krepcho, director of the *Second Harvest Food Bank* in Orlando has said that even college educated professional couples have begun to turn to food pantries.

By mid-2012, US food banks had expressed concerns on the expected difficulty in feeding the hungry over the coming months. Rapidly rising demand has been coinciding with higher food prices and with a decrease in donations, partly as the food industry is becoming more efficient and so has less mislabelled and other slightly defective food to give away. Also there has been less surplus federal food on offer. Additionally, there have been recent decreases in Federal funding, and Congress have been debating possible further cuts, including potentially billions of dollars from the Supplemental Nutrition Assistance Program (Food stamp programme). In September, Feeding America launched *Hunger Action Month*, with events planned all over the nation. Food banks and other agencies involved hope to raise awareness of the fact that about one in six Americans are struggling with hunger, and to get more Americans involved in helping out.

Europe

The first European food bank was founded in France during 1984. The first Italian food bank was established in 1989. Similar to the UK's experience, foodbanks have become much more common across continental Europe since the crisis that began in 2008, and especially since austerity began to take effect from late 2010.

In Spain, food banks can operate on the warehouse model, supplying a network of surrounding soup kitchens and other food relief agencies. The Spanish federation of food banks (FESBAL) helped to feed about 800,000 people during 2008-2011, according to the Carrefour Foundation. By October 2014, Spain had 55 food banks in total, with the numbers who depend on them having increased to 1.5 million.

In Belgium, food banks helped about 121,000 people during 2012. That was an increase of about 4,500 compared with 2011, the biggest increase since the start of the 2008 crisis. Belgian food banks account for about 65% of all food aid given out within the country.

The number of food banks has increased rapidly even in Germany, a country that has weathered the crisis relatively well, and has not needed to implement severe austerity. In 2012, professor Sabine Pfeiffer of Munich University of Applied Sciences said there has been an "explosion" of food bank usage.

The Most Deprived Persons Programme

While many European food banks are run by civil society with no government assistance, there is a project funded by the EU, the *Most deprived persons programme* (MDP), which specialises in supplying food to marginalised people who are not covered by the benefit system and who may be reluctant to approach the more formal food banks. Food is largely given out by Catholic churches. However, the EU is due to end funding for the MDP in 2013. Similar to the US, the EU no longer expects to need to buy much food to help farmers, as with high global food prices, market surpluses are now much less frequent, So there is less food available to hand out to food banks. In October 2012, the European Commission proposed a new Fund to replace the *Most deprived persons programme.*

United Kingdom

Though Foodbanks were rarely seen in the UK in the second half of the 20th century, their use has started to grow, especially in the 2000s, and have since dramatically expanded. The increase in the dependency on food banks has been blamed on the growing unemployment in the UK as well as the 2008 recession. The OECD found that people answering yes to the question 'Have there been times in the past 12 months when you did not have enough money to buy food that you or your family needed?' decreased from 9.8% in 2007 to 8.1% in 2012, leading some to say that the rise was due to both more awareness of food banks, and the government allowing Jobcentres to refer people to food banks when they were hungry (the previous Labour government had not allowed this).

As of January 2014 there were close to 1,000 UK food banks. The largest group co-ordinating UK foodbanks was The Trussell Trust, a Christian charity based in Salisbury. About 43% of the UK's foodbanks were run by Trussell, about 20% by smaller church networks such as Besom and Basics, about 31% were independent, and about 4% were run by secular food bank networks such as *Fare Share* and *Food Cycle.*

Before the financial crisis, food banks were "almost unheard of" in the UK. In 2004, Trussell only ran two food banks, but by 2007–2008, there were 22 food banks in the Trussell Trust Foodbank Network and by early 2011, The Trussell Trust supported 100. As of May 2012, they had 201. By August, 252. The rate of increase had been rising rapidly. In 2011, about one new food bank was being opened per week. In early 2012, about two were being opened each week. By July, The Trussell Trust had reported that the rate of new openings had increased to three per week. In August, the rate of new openings spiked at four per week, with three new food banks being opened in that month for Nottingham alone. By October 2012, the rate of increase had fallen back to about two or three per week.

Most UK food banks are hosted by churches in partnership with the wider community. They operate on the "frontline" model, giving out food directly to the hungry. Over 90% of the food given out is donated by the public, including schools, churches, businesses and individuals. The Trussell

Trust had aimed to provide short-term support for people whose needs have not yet been addressed by official state welfare provision; those who had been "falling into the cracks in the system". The Trussell franchise has procedures which aim to prevent long-term dependency on their services, and to ensure that those in need are referred to qualified outside agencies. The charity suggests that the *credit crunch* caused an upsurge in the number of people needing emergency food. Since 2010, demand for foodbanks continued to increase, and at a more rapid rate, partly as austerity began to take effect, and partly as those on low incomes began to draw down savings and run out of friends of whom they were willing to ask for help. Unlike *soup kitchens*, most UK food banks are unable to help people who come in off the street without a referral - instead they operate with a referral system. Vouchers are handed out to those in need by various sorts of frontline care professionals, such as social workers, health visitors, Citizens Advice Bureau, Jobcentres and housing officials. The voucher can typically be exchanged at the food bank for a package of food sufficient to last three days. The year to April 2013 saw close to 350,000 referrals to Trussel foodbanks, more than double the amount from the previous year.

A number of food banks have been set up outside of the Trussell system, in part as they don't like having to turn away people without referrals, although Trussell Trust foodbanks do help clients in need without vouchers to get one as quickly as possible. There is also FareShare, a London-based charity which operates some 19 depots on the US-style warehouse model. Rather than giving out food directly to individuals, FareShare distributes food to over 700 smaller agencies, mainly smaller independent operations like soup kitchens and breakfast clubs. Great emphasis is placed on reducing food waste as well as relieving food poverty. Fareshare operates on a business basis, employing a number of managers to oversee operations alongside their army of volunteers. Employee costs constituted over 50% of their expenditure in both 2011 and 2012. Additionally, charities receiving food do incur a charge for deliveries.

Another charity which operates on the US-style warehouse model and with a similar emphasis placed on reducing food waste as well as relieving food poverty is the Oxford Food Bank, which has a single base delivering to around 30 charities in the Oxford area. Although a much smaller enterprise than FareShare, it has a significantly lower pro-rata cost base as it employs no staff, with the whole operation up to director level run entirely by volunteers. This allows it to provide food at no cost to the recipient charities with all operating costs covered by grants and donations, supplying an estimated £25 of food at retail value for each £1 received in donations. In December 2012, it also started distributing food to needy families direct with the aid of local community centres and social services.

People who turn to food banks are typically grateful both for the food and for the warmth and kindness they receive from the volunteers. However, sometimes food banks have run out of supplies by the time they arrive. Some find it humiliating to have to ask for food, and that the packages they receive don't always seem nutritious. Some food banks have tried to respond with innovative programmes; *London Street Foodbank* for example has begun asking donors to send in supermarket vouchers so that those they serve will be able to choose food that best meets their nutritional needs.

Comparison to Other Countries

Food bank use in Germany and France is much higher than in Britain. In 2014, 1.5 million people a week used food banks in Germany and there are twice as many food banks in France as there are in Britain.

UK Food Bank Users

According to a May 2013 report by Oxfam and Church Action on Poverty, about half a million Britons had used food banks. The Trussel Trust reports that their food banks alone helped feed 346,992 people in 2012-13.

Close to half of those needing to use food banks have had issues with their benefits. Sanctioning benefits was the single most frequent reason for food bank referrals and there has been criticism over sanctions being imposed for allegedly spurious reasons.

A joint report from the Trussell Trust, the Church of England, and the charities Oxfam and Child Poverty Action Group found that food bank users were more likely to live in rented accommodation, be single adults or lone parents, be unemployed, and have experienced a "sanction," where their unemployment benefits were cut for at least one month.

Delay in paying housing benefit disability benefit and other benefits and general bureaucratic issues with benefits can force people to use food banks. Many further people who need food banks have low-income jobs, but struggle to buy food after making debt repayments and all expenses. Low paid workers, part-time workers and those with zero-hour contracts are particularly vulnerable to financial crisis and sometimes need food banks. As had been predicted, demand for food banks further increased after cuts to welfare came into effect in April 2013, which included the abolishment of Crisis loans. In April 2014 Trussell reported that they had handed out 913,000 food parcels in the last year, up from 347,000 the year before. Several councils have begun looking at funding food banks to increase their capability, as cuts to their budgets mean they will be less able to help vulnerable people directly.

The UK Government

Archbishop of Canterbury, Justin Welby urged the government to provide £150 million which he feels is needed to help prevent hunger in the UK. According to an all-party parliamentary report released in December 2014, key reasons for the increased demand for UK foodbanks are delays in paying benefits, welfare sanctions, and the recent reversal of the post WWII trend for poor people's incomes to rise above or in line with increased costs for housing, utility bills and food. However, the government has responded by saying that in fact the extra usage is likely due to more awareness of food banks as the proportion of benefits paid on time has risen from 88-89% under Labour, to 96-97% in 2014.

In 2013 the British government blocked a £22 million European Union fund to help finance food banks in the UK. This disappointed Labour MEP, Richard Howitt who assisted in negotiating the fund. Howlett stated:

It is very sad that our government is opposing this much-needed help for foodbanks on the basis that it is a national responsibility, when in reality it has no intention of providing the help itself. The only conclusion is that Conservative anti-European ideology is being put before the needs of the most destitute and deprived in our society.

Britain's prime Minister David Cameron said in the House of Commons in 2012 that he welcomed the efforts of food banks. Caroline Spelman, his Secretary of State for Environment, Food and

Rural Affairs, has described food banks as an "excellent example" of active citizenship. Labour MP Kate Green has a different view, feeling that the rise of food banks reflects people being let down by the state welfare system, saying "I feel a real burning anger about them ... People are very distressed at having to ask for food; it's humiliating and distressing." Cookery writer and poverty campaigner Jack Monroe wrote that those referred to food banks or given vouchers were "the lucky ones with a good doctor or health visitor who knows us well enough to recognise that something has gone seriously wrong" and expressed concern for those who lack this support.

Germany

There are over 900 food banks in Germany, up from just 1 in 1993. In 2014, 1.5 million people a week used food banks in Germany.

Asia

Several Asian countries have begun to use foodbanks; these include South Korea, Japan and Taiwan.

Singapore

In Singapore the Food Bank concept is being pioneered by the Food Bank Singapore.

India

Delhi food bank is an organisation that feeds, empowers and transforms lives in the New Delhi / NCR Region. They hold that their shared capabilities can make the basic aspiration of universal access to food a reality. They attempt to pursue this vision through high quality and standards for processes leveraged by technology to get the right aid to the right people at the right time.

Hong Kong

The first food bank in Hong Kong is Feeding Hong Kong. It was founded in 2009. Food Angel is also a food bank in Hong Kong.

Africa

The Egyptian Food Bank was established in Cairo in 2006, and less than ten years later, food banks run on similar principles spread to other Arab countries in North Africa and the Middle East.

In Sub-Saharan Africa, there are charity-run food banks that operate on a semi-commercial system that differs from both the more common "warehouse" and "frontline" models. In some rural LDCs such as Malawi, food is often relatively cheap and plentiful for the first few months after the harvest, but then becomes more and more expensive. Food banks in those areas can buy large amounts of food shortly after the harvest, and then as food prices start to rise, they sell it back to local people throughout the year at well below market prices. Such food banks will sometimes also act as centres to provide small holders and subsistence farmers with various forms of support.

Formed in 2009, *FoodBank South Africa* (FoodBank SA) is South Africa's national foodbanking network and a member of *The Global FoodBanking Network*. FoodBank SA's vision is "A South Africa without hunger and malnutrition".

Worldwide

Since the 1980s foodbanking has spread around the world. There are over 25 countries with active food bank groups under the umbrella of The Global FoodBanking Network. Countries in the international network include Australia, Israel, Turkey, Russia, India, Taiwan, Colombia, Brazil, Argentina, Chile, Guatemala, El Salvador, Nicaragua, Hong Kong, Singapore, South Korea and the UK. There are also several countries with foodbanks but which have not yet joined the network, either as they don't yet meet the required criteria or as they have not applied.

An alternative facility offering food to the hungry can be found worldwide wherever there are sizable Sikh communities. Long before foodbanks were invented, Langar has been making free vegetarian food available to Sikhs and non-Sikhs alike.

Reactions

Olivier De Schutter, a senior United Nations official, has cautioned Europe against allowing food banks to become a permanent partial replacement for welfare provision, as is the case in the U.S. and Canada.

The rise of food banks has been broadly welcomed. Not only do they provide a solution to the problem of hunger that doesn't require resources from the state, but they can be viewed as evidence of increasing community spirit and of active, caring citizenship. In the UK for example, Patrick Butler, society editor for The Guardian, has written that the rise of foodbanks has been most enthusiastically welcomed by the right, but also by many on the left of the political spectrum, who were often "nervously excited" about them. However, there has been considerable concern expressed by some researchers and politicians. Drawing on the United States's experience after the rapid rise of food banks in the 1980s, American sociology professor Janet Poppendieck warned that the rise of food banks can contribute to a long term erosion of human rights and support for entitlements. Once food banks become well established, it can be politically impossible to return responsibility for meeting the needs of hungry people to the state. Poppendieck says that the logistics of running food banks can be so demanding that they prevent kind hearted people from having time to participate in public policy advocacy; yet she also says if they can be encouraged to lobby politicians for long term changes that would help those on low income, they often have considerable credibility with legislators. As of 2012, senior US food banks workers have expressed a preference to remain politically neutral, which political activists have suggested may relate to their sources of funding.

Rachel Loopstra from University of Toronto has said foodbanks are often inefficient, unreliable and unable to supply nutritional food. She said a survey in Toronto found that only 1 in 5 families suffering from food insecurity would turn to food banks, in part as there is a stigma associated with having to do so. Elizabeth Dowler, Professor of Food & Social Policy at Warwick University, said that most British people prefer the state to take responsibility for helping the hungry. Hannah Lambie-Mumford, from Sheffield University, echoed the view that some users of food banks find having to ask for food humiliating, and also that food banks volunteers should be encouraged to advocate for long term solutions to the underlying causes of poverty and hunger.

Olivier De Schutter, a senior United Nations official charged with ensuring governments honour their obligation to safeguard their citizen's Right to food, has expressed alarm at the rise of food banks. He has reminded the governments of the advanced economies in Europe, Britain and Canada that they have a "duty to protect" their citizens from hunger, and suggested that leaving such an obligation to food banks may be an abuse of human rights.

Delhi Food Banking Network

Delhi Food Banking Network is a food bank system based in Delhi, India. Its goal is feeding and providing nutrition that helps break the cycle of poverty with linkages for healthcare, education, livelihoods for an integrated development objective catering to the vulnerable populations of Delhi and the National Capital Region.

The vulnerable populations being serviced by the Delhi FoodBank include feeding programs that serve school programs, charitable hospitals, orphanages, the destitute, beggars, homeless, faith based organisations and many others.

The Foodbank operates by matching food from donors to those that need this the most.

FoodBanking Technology

The Delhi Food Bank is leveraging mobile technologies through short code services to getfood and givefood.

Technology helps the management solutions for FoodBanking — specifically for process and workflow management — with the ability to adapt for sourcing, procurement, warehousing, delivery, monitoring, evaluation and training.

Through the sourcing process, donors are able to view the products that have been donated and view the inventory of those or despatch information to feeding programs. Historical transaction details are also available for a rapid reporting process. Recipients are able to view the details of donated products to their feeding programs and history of transactions. The warehouse managers are able to efficiently receive, manage the inventory and despatch through the automated processes.

The reporting process at each level of technology workflow enables users to track and generate customised reports based on their requirements. This also helps the monitoring, evaluation and learning by which feeding programs upload information of beneficiaries they have serviced.

India FoodBanking Network

In India the FoodBanking concept is being pioneered by the India FoodBanking Network to augment ongoing Government and Non Profit feeding programs, with the establishment of the Delhi FoodBank. IFBN acquires donated food in the form of grains, pulses, oil, spices etc. sourced on the basis of community food habits and makes these available through a network of community owned agencies. These institutional feeding programs include school feeding programs, shelter homes, Old age homes substance abuse clinics, after-school programs and cater to various needs.

India FoodBanking Network, of the Food Security Foundation India, is a national level project started under the leadership of Sam Pitroda, Advisor to the Prime Minister for Public Information and Infrastructure and Innovation, with the support of organisations like Cargill India, the DLF Foundation, the Reliance Foundation, the Global Alliance for Improved Nutrition and the Global FoodBanking Network, and is implemented and made operational by the Aidmatrix Foundation. IFBN works with a vision to support thousands of existing feeding programs and create a minimum one FoodBank in each district of India by 2020.

The Delhi Food Banking Network is a banking system that feeds, empowers and transforms lives in the New Delhi / NCR Region. We believe that our shared capabilities can make the most basic aspiration of universal access to food a reality. This is through high quality and standards for processes leveraged by technology to get the right aid to the right people at the right time.

Food Storage Container

Warehouse storing many kinds of food.

Food storage containers are widespread in use throughout the world and have probably been in use since the first human civilizations.

Consumer pantry, also diverse.

Early Civilizations

In early civilizations cereal grains such as maize, wheat, barley etc. were stored in large airy buildings, often raised up from the ground to reduce infestation by pests and vermin. Ancient Egyptian and early Hebrew writings include reference to such buildings and their successors can still be seen in use in less developed countries and regions.

Smaller quantities of food were stored in baskets made from woven grasses or leaves and such designs have remained in use to the present day.

Modern Containers

In the modern developed world, a very wide range of food packaging and containers is now available made from many materials.

Plastic containers:

Many products use low-density polyethylene formed into plastic bags or plastic boxes. There are many makers of plastic boxes some such as Tupperware and Lock&Lock are known throughout much of the developed world.

Metal containers:

Longer term storage or storage of items needing a higher degree of protection from the elements may use sheet metal. A common form of such storage is the biscuit tin.

Refrigerator:

Perhaps the most ubiquitous domestic item of food storage is the fridge or fridge-freezer in which a wide variety of foodstuffs are contained and preserved through the use of low temperatures.

Disposable:

Disposable foam food containers may be used for food or beverages.

Industrial Food Storage Containers

Wherever food is harvested, manufactured or distributed there is a need for containers to enable the food to travel securely and in good condition to the shop, warehouse or distribution depot. For many foods, especially those in their own individual containers such as canned vegetables, the common container is the corrugated fiberboard box, Corrugated boxes are available in many sizes.

Foam Food Container

A foam food container is a disposable container for various foods, including beverages. Food types may be processed (instant noodles), raw (meat from supermarkets), or cooked (delicatessen).

The containers are commonly used to serve takeout food from restaurants, and are also available by request for diners who wish to take the remainder of their meal home.

Siu mei with rice in a foam takeout container.

Khanom krok, Thai doughnuts with egg and coconut fillings in a foam container.

Construction

Foam take-out containers are made from EPS foam, or another type of polystyrene foam, and produced by injecting the foam into a mold. They are usually white in color, although they may be printed or impressed with a company logo or other message.

Styrofoam is a trademark of The Dow Chemical Company for closed-cell extruded polystyrene foam currently made for thermal insulation and craft applications. The term is often incorrectly used as a generic term for expanded (not extruded) polystyrene foam, such as disposable coffee cups, coolers, or cushioning material in packaging, which are typically white and are made of expanded polystyrene beads.

This is a different material from the extruded polystyrene used for Styrofoam insulation. The polystyrene foam used for craft applications, which can be identified by its roughness and by the fact that it "crunches" when cut, is moderately soluble in many organic solvents, cyanoacrylate, and the propellants and solvents of spray paint, and is not specifically identified as expanded or extruded. Another tradename for expanded polystyrene is thermacol, originated by BASF.

Varieties

Foam takeout containers come in different varieties. These may include:

- A rectangular shaped clamshell style container with an attached lid, which comes in various sizes. The larger ones usually feature several compartments, allowing different foods to be kept separated from each other. This variety usually features several small projections on the lid of the container, which fit through slots on the bottom half to "lock" it, therefore keeping the cover closed. These containers are leak-resistant only if kept upright, and often have a square imprint area for labeling.

- A cylindrical style container with a separate, translucent or opaque plastic lid which can seal tightly to resist leaks. The container may or may not taper somewhat towards the bottom. Both overall size, and the ratio of height to diameter can vary greatly. Such containers usually hold soups and stews; however, smaller varieties are often used to hold sauces and condiments.

Chinese Cuisine

This is the most commonly used takeout box for Chinese cuisines in East and Southeast Asia. It is standard for Cantonese cuisine in Hong Kong and many parts of China. It is sometimes used overseas in various restaurants, particularly in the United States and Canada.

Issues

Environmental Issues

Because foam takeout containers are entirely made out of polystyrene foam, these containers affect the environment as they do not biodegrade easily. However, Methanogenic consortia degrade styrene, using it as a carbon source. *Pseudomonas putida* can also convert styrene oil into various biodegradable polyhydroxyalkanoates. Some cities have gone as far as banning the use of foam take-out containers, notably San Francisco, Seattle and Portland, Oregon. In 2013, the mayor of New York City proposed banning foam food containers, for both health and environmental reasons. This plan began to be implemented in early 2015, however a September court ruling has since put the ban on hold.

Health Issues

It is debated whether styrene may migrate into food which is stored in foam food containers for even a small amount of time. Some researchers argue that polystyrene containers do pose a health risk, while industry defenders argue that trace amounts of styrene are already naturally present in food. Styrene foam containers can melt if the food or liquid is of a sufficient temperature. Some containers have been tested and labelled for safe use in microwave ovens; although the absence of such labeling does not mean a container is unsafe for this use, caution should still be taken.

Styrene is considered by both the EPA and IARC to be a possible carcinogen. It poses a health risk to workers during production of styrene and polystyrene items, with industries using a compliance program to deal with liabilities.Prolonged exposure to high amounts of styrene may affect the central nervous system, causing diseases like Parkinson's disease and multiple sclerosis.

Juicebox (Container)

A juice box, also called a carton or popper, is a small container used to conveniently carry and consume drinks (most often juice). They are frequently made of paperboard with an aluminum foil lining, but variations exist. Juice boxes are most popular with children, although other uses include emergency drinking water and wine.

A juice box.

Ruben Rausing first created a product in 1963 that consisted of a box that would be used for containing liquids, more specifically, milk. His creation was named the Tetra Brik, and gained popularity because the product was efficient and a major space saver compared to the canisters that were previously used. The juice box was officially incorporated in the U.S. market in 1980. After its introduction, the product gained almost instant popularity and the market began to grow at a fast rate. According to an article on the website E notes, in 1986, only six years after the product's introduction, juice boxes accounted for 20% of the United States juice market, as more and more companies were introducing their own lines of juice boxes.

Packaging and Marketing

Beetroot juice box packing, Germany.

A juice box is made of liquid packaging board which is usually six layers of paper, polyethylene, and aluminum foil. Paper is used to shape the product and give the box an extra source of strength, Polyethylene forms a liquid-tight seal and keeps the product dry, polyethylene is the layer used to print the information and graphics that are provided on the packaging, aluminum is used for the purpose of keeping light and oxygen out (as well preventing the juice from becoming spoiled without having to use extra preservatives). One of the reasons as to why juice boxes are so popular is their convenience. Juice boxes are portable and can be easily drunk while on-the-go. The shape of the product makes them easy for kids and adults to hold and use.

Juice boxes were originally designed in a very specific fashion, a style that has proved to be successful as it has remained, for the most part, unchanged. Manufacturers chose a box shape because they foresaw this shape as being the most convenient and easily handled. Juice boxes typically come with a covered hole and an attached bendable straw, which was introduced by Lynn Tilley, one of the designers who worked on porting it to America, after there were complaints about it being hard to drink from. These bendable straws make it easier for children to drink and generally results in less of a mess, and the bending nature of the straw helps it not fall into the box. However, there are some juice boxes available for purchase that are equipped with a pull tab; this variation of the juice box is ideal for larger portions that are not consumed in one sitting, because the tab is resealable.

At the juice box's inception in the early 1980s, juice boxes were primarily made up of diluted juice drinks containing many more artificial ingredients other than real fruit juice. However, as the product's popularity grew throughout the years, so did a preference with health and consuming natural products. So juice boxes evolved and began containing real fruit juice, as well as healthy add-ins, such as vitamins, to also appeal to a more health-conscious public.

Square Milk Jug

The square milk jug is a variant of the plastic gallon (3.785 litres) container of milk commonly sold in the United States . The design was introduced in the summer of 2008 and is marketed as environmentally friendly because of the shape's advantages for shipping and storage (better cube efficiency).

Square milk jugs are stackable and 9% more can be shipped in the same space. They also eliminate the need for milk crates, which are required for traditional bell-shaped jugs.

Pouring Technique

The technique used to pour milk from a square milk jug is different from that of a traditional milk jug. A Sam's Club representative suggests "tilt it slowly and pour slowly" and a dairy owner in Yerington, Nevada, where some square milk jugs are filled, described the pouring technique as "rock-and-pour instead of a lift-and-tip".

Complaints

Some consumers have complained that it is impossible to pour from this container without spilling, resulting in waste of time and milk, in addition to water and paper towels being used for cleanup. Social media has numerous complaints. Gains for the producer have resulted in a loss for the consumer.

As of fall, 2016, after about 8 years, this milk jug has been replaced with a design that works much more easily.

Design

Environmental benefits and cost savings are the primary stated benefits of square milk jugs, although there are noticeable trade-offs.

Because more milk fits on each truck, shipping costs can be reduced by half; this reduces the number of trucks on the road each year by 11000 (US estimate). Stores require five shipments per week in traditional containers now only require two, resulting in lower fuel consumption. Instead of being packed in reusable crates, the square jugs are shrink wrapped for shipment; however, the hundreds of square feet of single-use polyethylene film "stretch wrap" used on each pallet raises significant new environmental concerns. If one pallet uses approximately 175 linear feet of film, a 40 x 48 inch pallet 6 feet tall will hold 80 cubic feet or 360 gallons of milk (given 4.5 gallons of milk per cubic foot), and require 1,056 square feet of film, equaling just under 3 square feet of film per gallon. Only around 5% of stretch wrap is actually recycled; the remainder contributes to significant environmental fallout.

Manufacturers claim that this rectangular design has cut water use by 67% to 70%, according to one dairy, since there are no crates to be cleaned; the absence of dirty crates also reduces the risk of jugs carrying contamination. There is also the reduced cost with no crates to be stolen: Pennsylvania dairies alone spent US$6 million replacing milk crates in 2005, mostly from theft.

Impact

Critics of square milk jugs have noted increased spillage and difficulty pouring. This is a problem specific to the Gallon size that has a much larger mass when first opened.

There are also cost advantages resulting from lower shipping costs for the retailer and these can be passed on to the consumer. The cost to a consumer of a gallon of milk in a square milk jug is 10¢ to 20¢ cheaper than a gallon of milk in a traditional milk jug.

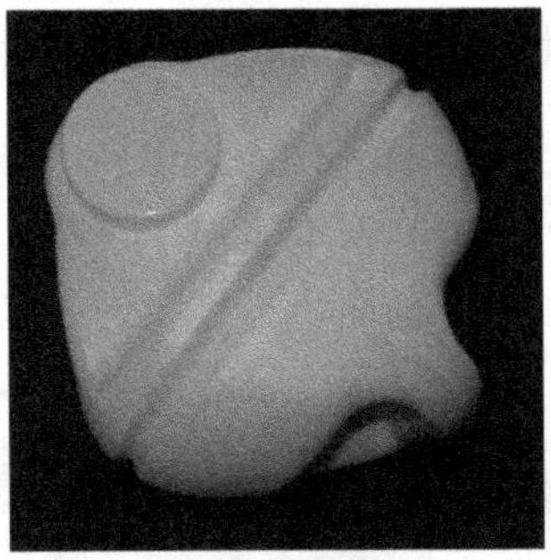

The top of a square milk jug. The ability to stack the square jugs was promoted as a cost saving improvement to the design.

Because square milk jugs can be stacked, they require less labor at the dairy, and can be packaged more quickly, which allows the milk to be shipped sooner, providing fresher milk to the retailer. The wide mouth is also easier to fill which reduces the likelihood of spillage and the sanitary issues caused by milk getting on the container.

The shape of the jug also allows more milk to be stored in a milk cooler. One Sam's Club store reported fitting 224 square gallon jugs in an area that only held 80 traditional one gallon jugs.

Reception

Square milk jugs have been adopted by some Wal-Mart, Costco and Sam's Club stores.

Consumers have criticized the square milk jug for being difficult to pour, especially for children.

When tilted shallowly, the larger opening (larger than a traditional milk jug), combined with its small lip, generates a wide stream of milk. This causes spills and leaks onto the sides of the container.

Some Sam's Club stores offered lessons on pouring, complete with complimentary cookies. Concern was also raised on the issue of the narrower base of the square jug, which allows it to be stored in the door of many refrigerators. Milk stored in this area may be more prone to spoiling due to frequent drastic changes in temperature.

Milk Bottle

Examples of milk bottles from the late 19th century made by the Warren Glass Works Company.

Milk bottles are bottles used for milk. They may be reusable glass bottles used mainly for doorstep delivery of fresh milk by milkmen. Customers are expected to rinse the empty bottles and leave on the doorstep for collection. The standard size of a bottle varies with location; common sizes are 1 pint or 1 quart, although cream may be delivered in smaller bottles.

More recently, plastic bottles have been commonly used for milk. These are often made of high-density polyethylene (HDPE), are used only once, and are easily recycleable.

History

A modern British milk bottle owned by Dairy Crest.

Before milk bottles, milkmen filled the customers' jugs. For many collectors, milk bottles carry a nostalgic quality of a bygone age. The most prized milk bottles are embossed or pyroglazed (painted) with names of dairies on them, which were used for home delivery of milk so that the milk bottles could find their way back to their respective dairies.

Seven modern Dairy Crest milk bottles.

It is not clear when the first milk bottles came into use. However, the New York Dairy Company is credited with having the first factory that produced milk bottles, and one of the first patents for a milk container was held by the Lester Milk Jar. There are many other similar milk containers from around this period, including the Mackworh Pure Jersey Cream crockery type jar, the Manorfield Stock Farm, the Manor, the Pa glass wide mouth jar, and the Tuthill's Dairy Unionville, NY.

Lewis P. Whiteman holds the first patent for a glass milk bottle with a small glass lid and a tin clip. The next earliest patent is for a milk bottle with a dome-type tin cap and was granted September 23, 1884 to Whitemen's brother, Abram V. Whiteman. This bottle has been found with cream line marks and is very valuable. The Whiteman brothers produced milk bottles based on these specifications at the Warren Glass Works Company in Cumberland, Maryland and sold them through their New York sales office.

The Original Thatcher is one of the most desirable milk bottles for collectors. The patent for the glass dome lid is dated April 27, 1886. There are several variations of this early milk bottle and many reproductions. During this time period, many types of bottles were being used to hold and distribute milk. These include a pop bottle type with a wire clamp, used by the Chicago Sterilized Milk Company, Sweet Clover, and others. Fruit jars were also used, but only the Cohansey Glass Manufacturing plant made them with dairy names embossed on them.

The Commonsense Milk Bottle with the first cap seat was developed as an economical means for sealing a reusable milk bottle by the Thatcher Manufacturing Company around 1900. Most bottles produced after this time have a cap seat.

By the 1920s, glass milk bottles had become the norm in the UK after slowly being introduced from the US before World War I.

Milk bottles before the 1930s were round. In 1935, slender-neck bottles were introduced in the UK. In the 1940s, a square squat bottle became the more popular style. Milk bottles since the 1930s have used pyroglaze or ACL (Applied Color Label) to identify the bottles. Before the 1930s, names were embossed on milk bottles using a slug plate. The name was impressed on the slug plate, then the plate was inserted into the mold used to make the bottle – the result was the embossed name on the bottle. In 1980 a new bottle, nicknamed "dumpy," was introduced in the UK where it remains the standard now.

From the 1960s onward in the United States, with improvements in shipping and storage materials, glass bottles have almost completely been replaced with either LDPE coated paper cartons or recyclable HDPE plastic containers (such as square milk jugs), depending on the brand. These paper and plastic containers are lighter, cheaper and safer to both manufacture and ship to consumers.

In 1975, 94% of milk in the UK was in glass bottles, but as of 2012 this number was down to 4%.

However, there are growing concerns among some Americans as to the quality and safety of industrialized milk, and the local, non-homogenized milk industry has seen a popular resurgence in certain markets in the US in the last decade or so. Because of this, the use of glass bottles in local or regional, non-industrial milk distribution has become an increasingly common sight.

Chronology

- 1878 – George Lester patents a glass milk container in New York.
- 1880 – British milk bottles were first produced by the Express Dairy Company. They were delivered by horse-drawn carts and delivered four times a day. The first bottles used a porcelain stopper top held on by wire. Lewis Whiteman patents the glass milk bottle with a glass lid.
- 1894 – Anthony Hailwood developed a pasteurisation process for milk which allowed it to be sterilized and be safely stored for longer periods. Milk could now be delivered once a day.
- 1920 – Advertisements began to appear on milk bottles. A sand-blasting technique was used to etch them on the glass.
- 1930s – Increased prevalence of battery electric vehicles as milk floats.
- mid 1950s – Cardboard tops were deemed unhygienic and banned in some locations. Delivery by horse-drawn carts was still common.
- early 1990s – The advertising largely disappeared with the introduction of infrared bottle scanners designed to check cleanliness.

Present Day

In some locations around the world, different colored tops on milk bottles indicate the fat content. Unpasteurized is often green-topped. However other colour designations may be used by some dairies. Bottles may also be marked, stamped or embossed with the name of the dairy.

In the United Kingdom, the aluminium foil tops on glass milk bottles are normally coloured:

Modern dairies may also use refillable plastic bottles, as well as plastic bottle tops. The colour coding for plastic milk bottle tops can be different from that of glass bottles. In the United Kingdom, the plastic tops on plastic milk bottles are normally coloured:

Colour	Meaning
Gold	Channel Island milk
Blue	Homogenized whole milk
Green	Semi-skimmed milk
Red	Skimmed milk
Orange	1% Fat milk

In the United Kingdom, milk sold to the door is mainly measured in Imperial pints (but labelled 568ml), because the glass bottles are 'returnable', which means they were excluded from metrication. However, Sterilized milk sold to the door is in 500ml glass bottles (and is 'non-returnable' and has colour coded lids that match the colour codes normally seen on plastic bottles). Often, in supermarkets they are sold in pints but labelled with their metric equivalent (568ml). Quantities larger than a pint are generally sold in metric units or multiples of a pint.

With lesser milk consumption, milk in Hong Kong is sold in both glass and plastic bottles as well as in cartons. The glass milk bottles are sold in supermarkets, convenience stores and in small restaurants. The glass milk bottle carries a deposit of 1 Hong Kong dollar. Most commonly available in the market in glass milk bottles are Kowloon Dairy and Trappist Dairy with different shaped glass milk bottles with both in school bottle size.

Orange juice and other fruit juices are also sold in doorstep deliveries in the same style of bottle used for milk. Typically these have an aluminium foil top colour coded to indicate the flavour.

In some countries (e.g. Estonia and some provinces of Canada), it is common to buy milk in a milk bag.

While the proportion of sales in milk bags in the United Kingdom is quite low, semi-skimmed milk is sold in bags by Dairy Crest/Milk and More, and in Sainsbury's supermarket it is available in Whole milk, Semi-skimmed milk and Skimmed milk options. However supermarket availability is limited to larger branches.

Refrigerator

Food in a refrigerator with its door open.

A side-by-side refrigerator-freezer with an icemaker.

A refrigerator (colloquially fridge) is a popular household appliance that consists of a thermally insulated compartment and a heat pump (mechanical, electronic or chemical) that transfers heat from the inside of the fridge to its external environment so that the inside of the fridge is cooled to a temperature below the ambient temperature of the room. Refrigeration is an essential food storage technique in developed countries. The lower temperature lowers the reproduction rate of bacteria, so the refrigerator reduces the rate of spoilage. A refrigerator maintains a temperature a few degrees above the freezing point of water. Optimum temperature range for perishable food storage is 3 to 5 °C (37 to 41 °F). A similar device that maintains a temperature below the freezing point of water is called a freezer. The refrigerator replaced the icebox, which had been a common household appliance for almost a century and a half. For this reason, a refrigerator is sometimes referred to as an icebox in American usage.

The first cooling systems for food involved using ice. Artificial refrigeration began in the mid-1750s, and developed in the early 1800s. In 1834, the first working vapor-compression refrigeration system was built. The first commercial ice-making machine was invented in 1854. In 1913, refrigerators for home use were invented. In 1923 Frigidaire introduced the first self-contained unit. The introduction of Freon in the 1920s expanded the refrigerator market during the 1930s. Home freezers as separate compartments (larger than necessary just for ice cubes) were introduced in 1940. Frozen foods, previously a luxury item, became commonplace.

Freezer units are used in households and in industry and commerce. Commercial refrigerator and freezer units were in use for almost 40 years prior to the common home models. Most households use the freezer-on-top-and-refrigerator-on-bottom style, which has been the basic style since the 1940s. A vapor compression cycle is used in most household refrigerators, refrigerator–freezers and freezers. Newer refrigerators may include automatic defrosting, chilled water and ice from a dispenser in the door.

Domestic refrigerators and freezers for food storage are made in a range of sizes. Among the smallest is a 4 L Peltier refrigerator advertised as being able to hold 6 cans of beer. A large domestic refrigerator stands as tall as a person and may be about 1 m wide with a capacity of 600 L. Refrigerators and freezers may be free-standing, or built into a kitchen. The refrigerator allows the modern family to keep food fresh for longer than before. Freezers allow people to buy food in bulk and eat it at leisure, and bulk purchases save money.

History

Refrigeration Technology

Before the invention of the refrigerator, icehouses were used to provide cool storage for most of the year. Placed near freshwater lakes or packed with snow and ice during the winter, they were once very common. Natural means are still used to cool foods today. On mountainsides, runoff from melting snow is a convenient way to cool drinks, and during the winter one can keep milk fresh much longer just by keeping it outdoors. The word "refrigeratory" was used at least as early as the 17th century.

The history of artificial refrigeration began when Scottish professor William Cullen designed a small refrigerating machine in 1755. Cullen used a pump to create a partial vacuum over a container of diethyl ether, which then boiled, absorbing heat from the surrounding air. The experiment even created a small amount of ice, but had no practical application at that time.

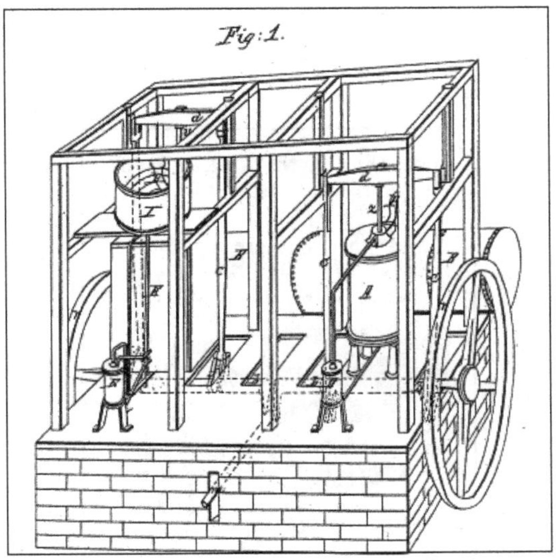

Schematic of Dr. John Gorrie's 1841 mechanical ice machine.

In 1805, American inventor Oliver Evans described a closed vapor-compression refrigeration cycle for the production of ice by ether under vacuum. In 1820, the British scientist Michael Faraday liquefied ammonia and other gases by using high pressures and low temperatures, and in 1834, an American expatriate to Great Britain, Jacob Perkins, built the first working vapor-compression refrigeration system. It was a closed-cycle device that could operate continuously. A similar attempt was made in 1842, by American physician, John Gorrie, who built a working prototype, but it was a commercial failure. American engineer Alexander Twining took out a British patent in 1850 for a vapor compression system that used ether.

The first practical vapor compression refrigeration system was built by James Harrison, a British journalist who had emigrated to Australia. His 1856 patent was for a vapor compression system using ether, alcohol or ammonia. He built a mechanical ice-making machine in 1851 on the banks of the Barwon River at Rocky Point in Geelong, Victoria, and his first commercial ice-making machine followed in 1854. Harrison also introduced commercial vapor-compression refrigeration to breweries and meat packing houses, and by 1861, a dozen of his systems were in operation.

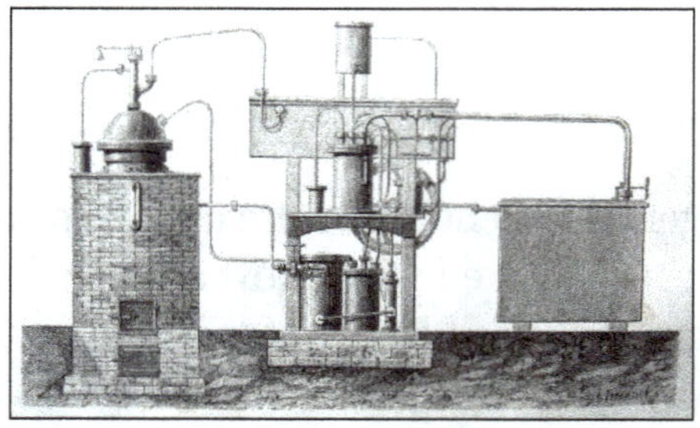

Ferdinand Carré's ice-making device.

The first gas absorption refrigeration system using gaseous ammonia dissolved in water (referred to as "aqua ammonia") was developed by Ferdinand Carré of France in 1859 and patented in 1860. Carl von Linde, an engineering professor at the Technological University Munich in Germany, patented an improved method of liquefying gases in 1876. His new process made possible the use of gases such as ammonia, sulfur dioxide (SO_2) and methyl chloride (CH_3Cl) as refrigerants and they were widely used for that purpose until the late 1920s.

Domestic Refrigerator

McCray pre-electric home refrigerator ad (1905) This company, founded in 1887, is still in business.

In 1913, refrigerators for home and domestic use were invented by Fred W. Wolf of Fort Wayne, Indiana with models consisting of a unit that was mounted on top of an ice box. In 1914, engineer Nathaniel B. Wales of Detroit, Michigan, introduced an idea for a practical electric refrigeration unit, which later became the basis for the Kelvinator. A self-contained refrigerator, with a compressor on the bottom of the cabinet was invented by Alfred Mellowes in 1916. Mellowes produced this refrigerator commercially but was bought out by William C. Durant in 1918, who started the Frigidaire Company to mass-produce refrigerators. In 1918, Kelvinator Company introduced the first refrigerator with any type of automatic control. The absorption refrigerator was invented by Baltzar von Platen and Carl Munters from Sweden in 1922, while they were still students at the Royal Institute of Technology in Stockholm. It became a worldwide success and was commercialized by Electrolux. Other pioneers included Charles Tellier, David Boyle, and Raoul Pictet. Carl von Linde was the first to patent and make a practical and compact refrigerator.

These home units usually required the installation of the mechanical parts, motor and compressor, in the basement or an adjacent room while the cold box was located in the kitchen. There was a 1922 model that consisted of a wooden cold box, water-cooled compressor, an ice cube tray and a 9-cubic-foot (0.25 m³) compartment, and cost $714. (A 1922 Model-T Ford cost about $450.) By 1923, Kelvinator held 80 percent of the market for electric refrigerators. Also in 1923 Frigidaire introduced the first self-contained unit. About this same time porcelain-covered metal cabinets began to appear. Ice cube trays were introduced more and more during the 1920s; up to this time freezing was not an auxiliary function of the modern refrigerator.

General Electric "Monitor-Top" refrigerator, introduced in 1927.

The first refrigerator to see widespread use was the General Electric "Monitor-Top" refrigerator introduced in 1927, so-called because of its resemblance to the gun turret on the ironclad warship USS *Monitor* of the 1860s. The compressor assembly, which emitted a great deal of heat, was placed above the cabinet, and enclosed by a decorative ring. Over a million units were produced. As the refrigerating medium, these refrigerators used either sulfur dioxide, which is corrosive to the eyes and may cause loss of vision, painful skin burns and lesions, or methyl formate, which is highly flammable, harmful to the eyes, and toxic if inhaled or ingested. Many of these units are still functional today, after requiring little more service than a replacement start relay or thermostat if at all. These cooling systems cannot legally be recharged with the hazardous original refrigerants if they leak or break down.

The introduction of Freon in the 1920s expanded the refrigerator market during the 1930s and provided a safer, low-toxicity alternative to previously used refrigerants. Separate freezers became common during the 1940s; the popular term at the time for the unit was a *deep freeze*. These devices, or *appliances*, did not go into mass production for use in the home until after World War II. The 1950s and 1960s saw technical advances like automatic defrosting and automatic ice making. More efficient refrigerators were developed in the 1970s and 1980s, even though environmental issues led to the banning of very effective (Freon) refrigerants. Early refrigerator models (from 1916) had a cold compartment for ice cube trays. From the late 1920s fresh vegetables were successfully processed through freezing by the Postum Company (the forerunner of General Foods), which had acquired the technology when it bought the rights to Clarence Birdseye's successful fresh freezing methods.

The first successful application of frozen foods occurred when General Foods heiress Marjorie Merriweather Post (then wife of Joseph E. Davies, United States Ambassador to the Soviet Union) deployed commercial-grade freezers in Spaso House, the US Embassy in Moscow, in advance of the Davies' arrival. Post, fearful of the USSR's food processing safety standards, fully stocked the freezers with products from General Foods' Birdseye unit. The frozen food stores allowed the Davies to entertain lavishly and serve fresh frozen foods that would otherwise be out of season. Upon returning from Moscow, Post (who resumed her maiden name after divorcing Davies) directed General Foods to market frozen product to upscale restaurants.

Home freezers as separate compartments (larger than necessary just for ice cubes), or as separate units, were introduced in the United States in 1940. Frozen foods, previously a luxury item, became commonplace.

Freezer

Freezer units are used in households and in industry and commerce. Food stored at or below −18 °C (0 °F) is safe indefinitely. Most household freezers maintain temperatures from −23 to −18 °C (−9 to 0 °F), although some freezer-only units can achieve −34 °C (−29 °F) and lower. Refrigerators generally do not achieve lower than −23 °C (−9 °F), since the same coolant loop serves both compartments: Lowering the freezer compartment temperature excessively causes difficulties in maintaining above-freezing temperature in the refrigerator compartment. Domestic freezers can be included as a separate compartment in a refrigerator, or can be a separate appliance. Domestic freezers are generally upright units resembling refrigerators or chests (upright units laid on their backs). Many modern upright freezers come with an ice dispenser built into their door. Some upscale models include thermostat displays and controls, and sometimes flatscreen televisions as well.

Commercial and Domestic Refrigerators

Commercial refrigerator and freezer units, which go by many other names, were in use for almost 40 years prior to the common home models. They used gas systems such as ammonia (R-717) or sulfur dioxide (R-764), which occasionally leaked, making them unsafe for home use. Practical household refrigerators were introduced in 1915 and gained wider acceptance in the United States in the 1930s as prices fell and non-toxic, non-flammable synthetic refrigerants such as Freon-12 (R-12) were introduced. However, R-12 damaged the ozone layer, causing governments to issue a ban on its use in new refrigerators and air-conditioning systems in 1994. The less harmful replacement for R-12, R-134a (tetrafluoroethane), has been in common use since 1990, but R-12 is still found in many old systems today.

A common commercial refrigerator is the glass fronted beverage cooler. These type of appliances are typically designed for specific re-load conditions meaning that they generally have a larger cooling system. This ensures that they are able to cope with a large throughput of drinks and frequent door opening. As a result, it is common for these types of commercial refrigerators to have energy consumption of >4 kWh/day.

Styles of Refrigerators

In the early 1950s most refrigerators were white, but from the mid-1950s through present day designers and manufacturers put color onto refrigerators. In the late-1950s/early-1960s, pastel colors like turquoise and pink became popular, brushed chrome-plating (similar to stainless finish) was available on some models from different brands. In the late 1960s and throughout the 1970s, earth tone colors were popular, including Harvest Gold, Avocado Green and almond. In the 1980s, black became fashionable. In the late 1990s stainless steel came into vogue, and in 2009, one manufacturer introduced multi-color designs.

Frigidaire Imperial "Frost Proof" model FPI-16BC-63, top refrigerator/bottom freezer with brushed chrome door finish made by General Motors Canada in 1963.

Production by Country

Country	Production	Year
China	29,871,000	(2005)
United States	11,639,000	(2003)
Italy	7,201,000	(2004)
South Korea	7,122,000	(2004)
Turkey	4,867,000	(2003)
India	3,715,000	(2003)
Brazil	3,544,000	(2003)
Japan	2,821,000	(2005)
Mexico	2,291,000	(2004)
Thailand	2,246,000	(1996)
Germany	2,061,000	(2004)
Hungary	1,625,000	(2004)
Poland	1,618,000	(2005)
Spain	1,269,000	(1995)
Romania	1,169,000	(2005)
Belarus	995,000	(2005)
Slovenia	863,000	(1995)
Egypt	808,000	(2003)
United Kingdom	745,000	(2003)

Country	Production	Year
South Africa	711,000	(2003)
Sweden	639,000	(2004)
Ukraine	562,000	(1995)
France	544,000	(2003)
Australia	423,000	(1995)
Portugal	399,000	(2004)
Bulgaria	353,000	(2005)
Slovakia	330,000	(1995)
Indonesia	291,000	(1995)
Malaysia	187,000	(2003)
Algeria	150,000	(2003)
Lithuania	107,000	(2004)
Finland	104,000	(1995)
Argentina	49,000	(1995)
Republic of Moldova	24,300	(1995)
Uzbekistan	18,600	(1995)
Azerbaijan	13,400	(2005)
Kazakhstan	10,900	(1995)
Tajikistan	50	(1995)

General Technical Explanation

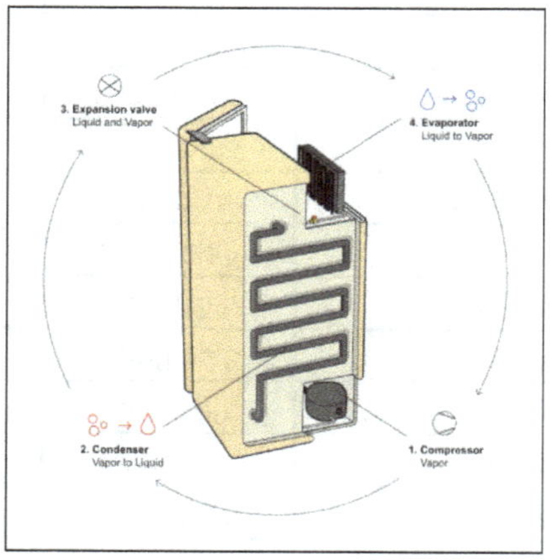

Basic functioning of a refrigerator.

A vapor compression cycle is used in most household refrigerators, refrigerator–freezers and freezers. In this cycle, a circulating refrigerant such as R134a enters a compressor as low-pressure vapor at or slightly below the temperature of the refrigerator interior. The vapor is compressed and exits the compressor as high-pressure superheated vapor. The superheated vapor travels under pressure through coils or tubes that make up the *condenser*; the coils or tubes are passively cooled by exposure to air in the room. The condenser cools the vapor, which liquefies. As the refrigerant leaves the condenser, it is still under pressure but is now only slightly above room temperature. This liquid refrigerant is forced through a metering or throttling device, also known as an expansion valve (essentially a pin-hole sized constriction in the tubing) to an area of much lower pressure. The sudden decrease in pressure results in explosive-like flash evaporation of a portion (typically about half) of the liquid. The latent heat absorbed by this flash evaporation is drawn mostly from adjacent still-liquid refrigerant, a phenomenon known as *auto-refrigeration*. This cold and partially vaporized refrigerant continues through the coils or tubes of the evaporator unit. A fan blows air from the refrigerator or freezer compartment ("box air") across these coils or tubes and the refrigerant completely vaporizes, drawing further latent heat from the box air. This cooled air is returned to the refrigerator or freezer compartment, and so keeps the box air cold. Note that the cool air in the refrigerator or freezer is still warmer than the refrigerant in the evaporator. Refrigerant leaves the evaporator, now fully vaporized and slightly heated, and returns to the compressor inlet to continue the cycle.

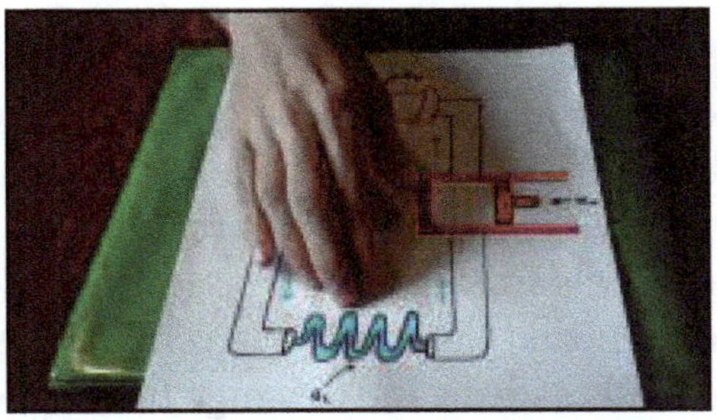

Process and Components of a Conventional Refrigerator.

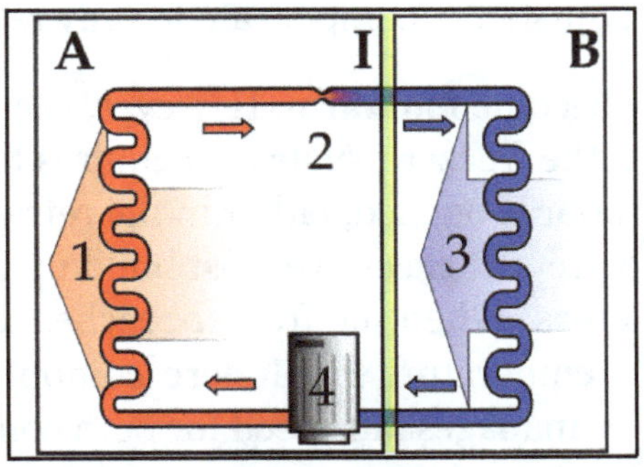

Vapor Compression Cycle – A: hot compartment (kitchen), B: cold compartment (refrigerator box), I: insulation, 1: Condenser, 2: Expansion valve, 3: Evaporator unit, 4: Compressor.

An Embraco compressor and fan-assisted condenser coil.

Domestic refrigerators are extremely reliable because the moving parts and fluids are sealed from the atmosphere for life, with no possibility of leakage or contamination. In comparison, mechanically-driven refrigeration compressors, such as those in automobile air conditioning, inevitably leak fluid and lubricant past the shaft seals. This leads to a requirement for periodic recharging and, if ignored, possible compressor failure.

An absorption refrigerator works differently from a compressor refrigerator, using a source of heat, such as combustion of liquefied petroleum gas, solar thermal energy or an electric heating element. These heat sources are much quieter than the compressor motor in a typical refrigerator. A fan or pump might be the only mechanical moving parts; reliance on convection is considered impractical.

The Peltier effect uses electricity to pump heat directly; refrigerators employing this system are sometimes used for camping, or in situations where noise is not acceptable. They can be totally silent (if a fan for air circulation is not fitted) but are less energy-efficient than other methods.

Other uses of an absorption refrigerator (or "chiller") include large systems used in office buildings or complexes such as hospitals and universities. These large systems are used to chill a brine solution that is circulated through the building.

Many modern refrigerator/freezers have the freezer on top and the refrigerator on the bottom. Most refrigerator-freezers—except for manual defrost models or cheaper units—use what appears to be two thermostats. Only the refrigerator compartment is properly temperature controlled. When the refrigerator gets too warm, the thermostat starts the cooling process and a fan circulates the air around the freezer. During this time, the refrigerator also gets colder. The freezer control knob only controls the amount of air that flows into the refrigerator via a damper system. Changing the refrigerator temperature will inadvertently change the freezer temperature in the opposite direction. Changing the freezer temperature will have no effect on the refrigerator temperature. The freezer control may also be adjusted to compensate for any refrigerator adjustment.

This means the refrigerator may become too warm. However, because only enough air is diverted to the refrigerator compartment, the freezer usually re-acquires the set temperature quickly, unless the door is opened. When a door is opened, either in the refrigerator or the freezer, the fan in some units stops immediately to prevent excessive frost build up on the freezer's evaporator coil, because this coil is cooling two areas. When the freezer reaches temperature, the unit cycles off, no matter what the refrigerator temperature is. Modern computerized refrigerators do not use the damper system. The computer manages fan speed for both compartments, although air is still blown from the freezer.

A few manufacturers offer dual compressor models. These models have separate freezer and refrigerator compartments that operate independently of each other, sometimes mounted within a single cabinet. Each has its own separate compressor, condenser and evaporator coils, insulation, thermostat, and door. Typically, the compressors and condenser coils are mounted at the top of the cabinet, with a single fan to cool them both.

This design, where no air passes between the two compartments, provides for more appropriate humidity levels and much tighter temperature control in each compartment. It also requires much less energy to operate, since each compressor & coolant system can be optimized for a specific temperature range. Further, opening the door of one compartment does not affect the temperature of the air or humidity level in the other compartment. Thus, it avoids many of the disadvantages of the much more common single compressor designs described above, although at a higher initial cost and increased system noise. Manufacturers of such designs argue that the increased cost is compensated over time due to reduced energy use and less food waste due to reduced spoilage.

Alternatives to the vapor-compression cycle not in current use include;

- Acoustic cooling
- Air cycle
- Magnetic cooling
- Malone engine
- Pulse tube
- Stirling cycle
- Thermoelectric cooling and thermionic cooling
- Vortex tube
- Water cycle systems

Features

The inside of a home refrigerator containing a large variety of everyday food items.

Newer refrigerators may include:

- Automatic defrosting.
- A power failure warning that alerts the user by flashing a temperature display. It may display the maximum temperature reached during the power failure, and whether frozen food has defrosted or may contain harmful bacteria.
- Chilled water and ice from a dispenser in the door. Water and ice dispensing became available in the 1970s. In some refrigerators, the process of making ice is built-in so the user doesn't have to manually use ice trays. Some refrigerators have water chillers and water filtration systems.
- Cabinet rollers that lets the refrigerator roll out for easier cleaning.
- Adjustable shelves and trays.
- A status indicator that notifies when it is time to change the water filter.

- An in-door ice caddy, which relocates the ice-maker storage to the freezer door and saves approximately 60 litres (2 cu ft) of usable freezer space. It is also removable, and helps to prevent ice-maker clogging.
- A cooling zone in the refrigerator door shelves. Air from the freezer section is diverted to the refrigerator door, to cool milk or juice stored in the door shelf.
- A drop down door built into the refrigerator main door, giving easy access to frequently used items such as milk, thus saving energy by not having to open the main door.
- A Fast Freeze function to rapidly cool foods by running the compressor for a predetermined amount of time and thus temporarily lowering the freezer temperature below normal operating levels. It is recommended to use this feature several hours before adding more than 1 kg of unfrozen food to the freezer. For freezers without this feature, lowering the temperature setting to the coldest will have the same effect.

Early freezer units accumulated ice crystals around the freezing units. This was a result of humidity introduced into the units when the doors to the freezer were opened condensing on the cold parts, then freezing. This frost buildup required periodic thawing ("defrosting") of the units to maintain their efficiency. Manual Defrost (referred to as Cyclic) units are still available. Advances in automatic defrosting eliminating the thawing task were introduced in the 1950s, but are not universal, due to energy performance and cost. These units used a counter that only defrosted the freezer compartment (Freezer Chest) when a specific number of door openings had been made. The units were just a small timer combined with an electrical heater wire that heated the freezer's walls for a short amount of time to remove all traces of frost/frosting. Also, early units featured freezer compartments located within the larger refrigerator, and accessed by opening the refrigerator door, and then the smaller internal freezer door; units featuring an entirely separate freezer compartment were introduced in the early 1960s, becoming the industry standard by the middle of that decade. These older freezer compartments were the main cooling body of the refrigerator, and only maintained a temperature of around −6 °C (21 °F), which is suitable for keeping food for a week.

In the early 1950s, the butter conditioner's patent was filed and published by the inventor Nave Alfred E. This feature was supposed to "provide a new and improved food storage receptacle for storing butter or the like which may quickly and easily be removed from the refrigerator cabinet for the purpose of cleaning." Because of the high interest to the invention, companies in UK, New Zealand, and Australia started to include the feature into the mass fridge production and soon it became a symbol of the local culture. However, not long after that it was removed from production as according to the companies this was the only way for them to meet new ecology regulations and they found it inefficient to have a heat generating device inside a fridge.

Later advances included automatic ice units and self compartmentalized freezing units.

An increasingly important environmental concern is the disposal of old refrigerators— initially because freon coolant damages the ozone layer—but as older generation refrigerators wear out, the destruction of CFC-bearing insulation also causes concern. Modern refrigerators usually use a refrigerant called HFC-134a (1,1,1,2-Tetrafluoroethane), which does not deplete the ozone layer, instead of Freon. A R-134a is now becoming very uncommon in Europe. Newer refrigerants are being used instead. The main refrigerant now used is R-600a, or isobutane. This refrigerant is naturally occurring, and has a smaller effect on the atmosphere if released. There have been reports of refrigerators exploding if the refrigerant leaks gas in the presence of a spark.

Disposal of discarded refrigerators is regulated, often mandating the removal of doors; children playing hide-and-seek have been asphyxiated while hiding inside discarded refrigerators, particularly older models with latching doors. Since 2 August 1956, under U.S. federal law, refrigerator doors are no longer permitted to latch so they cannot be opened from the inside. Modern units use a magnetic door gasket that holds the door sealed but allows it to be pushed open from the inside. This gasket was invented, developed and manufactured by Max Baermann (1903-1984) of Bergisch Gladbach/Germany.

Types of Domestic Refrigerators

Domestic refrigerators and freezers for food storage are made in a range of sizes. Among the smallest is a 4 L Peltier refrigerator advertised as being able to hold 6 cans of beer. A large domestic refrigerator stands as tall as a person and may be about 1 m wide with a capacity of 600 L. Some models for small households fit under kitchen work surfaces, usually about 86 cm high. Refrigerators may be combined with freezers, either stacked with refrigerator or freezer above, below, or side by side. A refrigerator without a frozen food storage compartment may have a small section just to make ice cubes. Freezers may have drawers to store food in, or they may have no divisions (chest freezers).

Refrigerators and freezers may be free-standing, or built into a kitchen.

- Compressor refrigerators are by far the most common type; they make a noticeable noise.
- Absorption refrigerators or thermo-electric Peltier units are used where quiet running is required; Peltier coolers are used in the smallest refrigerators as they have no bulky mechanism.
- Compressor and Peltier refrigerators are powered by electricity. Absorption units can be designed to get power from any heat source. A noticeable difference between the two types is the absence of refrigerant with Peltier coolers (these use a different method of cooling). But Peltier coolers use more electricity because they are thermodynamically inefficient.
- Oil, gas (natural gas or propane) and dual power gas/electricity units are also available (typically found in RV's).
- Solar refrigerators and Thermal mass refrigerators are designed to reduce electrical consumption. Solar refrigerators have the added advantage that they do not use refrigerants that are harmful to the environment or flammable. Typical solar designs are absorption refrigerators that use ammonia as the working gas, and employ large mirrors to concentrate sufficient sunlight to reach the temperature required to free gaseous ammonia from the solvent. Most thermal mass refrigerators are designed to use electricity intermittently. As these units are heavily insulated, cooling load is limited primarily to heat introduced by new items to be refrigerated, and ambient air transfer when the unit is open. Very little power is therefore required if opened infrequently. Refrigeration units for commercial and industrial applications can be made in various size, shape or style to fit customer needs.

Other specialised cooling mechanisms may be used for cooling, but have not been applied to domestic refrigerators.

- Magnetic refrigerators are refrigerators that work on the magnetocaloric effect. The cooling effect is triggered by placing a metal alloy in a magnetic field.
- Acoustic refrigerators are refrigerators that use resonant linear reciprocating motors/alternators to generate a sound that is converted to heat and cold using compressed helium gas. The heat is discarded and the cold is routed to the refrigerator.

Energy Efficiency

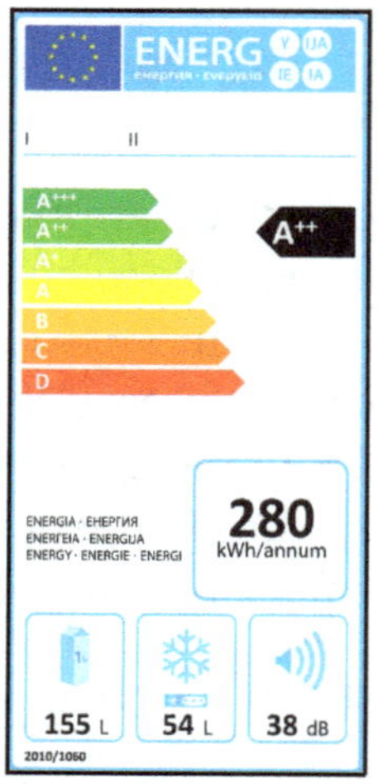

European energy label for a fridge.

In a house without air-conditioning (space heating and/or cooling) refrigerators consumed more energy than any other home device. In the early 1990s a competition was held among the major manufacturers to encourage energy efficiency. Current US models that are Energy Star qualified use 50% less energy than the average models made in 1974. The most energy-efficient unit made in the US consumes about half a kilowatt-hour per day (equivalent to 20 W continuously). But even ordinary units are quite efficient; some smaller units use less than 0.2 kWh per day (equivalent to 8 W continuously). Larger units, especially those with large freezers and icemakers, may use as much as 4 kW·h per day (equivalent to 170 W continuously). The European Union uses a letter-based mandatory energy efficiency rating label instead of the Energy Star; thus EU refrigerators at the point of sale are labelled according to how energy-efficient they are.

Modern refrigerators.

For US refrigerators, the Consortium on Energy Efficiency (CEE) further differentiates between Energy Star qualified refrigerators. Tier 1 refrigerators are those that are 20% to 24.9% more efficient than the Federal minimum standards set by the National Appliance Energy Conservation Act (NAECA). Tier 2 are those that are 25% to 29.9% more efficient. Tier 3 is the highest qualification, for those refrigerators that are at least 30% more efficient than Federal standards. About 82% of the Energy Star qualified refrigerators are Tier 1, with 13% qualifying as Tier 2, and just 5% at Tier 3.

Besides the standard style of compressor refrigeration used in normal household refrigerators and freezers, there are technologies such as absorption refrigeration and magnetic refrigeration. Although these designs generally use a much larger amount of energy compared to compressor refrigeration, other qualities such as silent operation or the ability to use gas can favor these refrigeration units in small enclosures, a mobile environment or in environments where unit failure would lead to devastating consequences.

Many refrigerators made in the 1930s and 1940s were far more efficient than most that were made later. This is partly attributable to the addition of new features, such as auto-defrost, that reduced efficiency. Additionally,after World War 2, refrigerator style became more important than efficiency. This was especially true in the 1970s, when side-by-side models with ice dispensers and water chillers became popular. However, the reduction in efficiency also arose partly from reduction in the amount of insulation to cut costs. Because of the introduction of new energy efficiency standards, refrigerators made today are much more efficient than those made in the 1930s; they consume the same amount of energy while being three times as large.

The efficiency of older refrigerators can be improved by defrosting (if the unit is manual defrost) and cleaning them regularly, replacing old and worn door seals with new ones, adjusting the thermostat to accommodate the actual contents (a refrigerator needn't be colder than 4 °C (39 °F) to store drinks and non-perishable items) and also replacing insulation, where applicable. Some sites recommend you clean condenser coils every month or so on units with coils on the rear. It has been proven that this does very little for improving efficiency, however, the unit should be able to "breathe" with adequate spaces around the front, back, sides and above the unit. If the refrigerator uses a fan to keep the condenser cool, then this must be cleaned, at the very least, yearly.

Frost-free refrigerators or freezers use electric fans to cool the appropriate compartment. This could be called a "fan forced" refrigerator, whereas manual defrost units rely on colder air lying at the bottom, versus the warm air at the top to achieve adequate cooling. The air is drawn in through an inlet duct and passed through the evaporator where it is cooled, the air is then circulated throughout the cabinet via a series of ducts and vents. Because the air passing the evaporator is supposedly warm and moist, frost begins to form on the evaporator (especially on a freezer's evaporator). In cheaper and/or older models, a defrost cycle is controlled via a mechanical timer. This timer is set to shut off the compressor and fan and energize a heating element located near or around the evaporator for about 15 to 30 minutes at every 6 to 12 hours. This melts any frost or ice build up and allows the refrigerator to work normally once more. It is believed that frost free units have a lower tolerance for frost, due to their air-conditioner like evaporator coils. Therefore, if a door is left open accidentally (especially the freezer), the defrost system may not remove all frost, in this case, the freezer (or refrigerator) must be defrosted.

If the defrosting system melts all the ice before the timed defrosting period ends, then a small device (called a defrost limiter) acts like a thermostat and shuts off the heating element to prevent too large a temperature fluctuation, it also prevents hot blasts of air when the system starts again, should it finish defrosting early. On some early frost-free models, the defrost limiter also sends a signal to the defrost timer to start the compressor and fan as soon as it shuts off the heating element before the timed defrost cycle ends. When the defrost cycle is completed, the compressor and fan are allowed to cycle back on.

Frost-free refrigerators, including some early frost free refrigerator/freezers that used a cold plate in their refrigerator section instead of airflow from the freezer section, generally don't shut off their refrigerator fans during defrosting. This allows consumers to leave food in the main refrigerator compartment uncovered, and also helps keep vegetables moist. This method also helps reduce energy consumption, because the refrigerator is above freeze point and can pass the warmer-than-freezing air through the evaporator or cold plate to aid the defrosting cycle.

Regarding total life-cycle costs, many governments offer incentives to encourage recycling of old refrigerators. One example is the Phoenix refrigerator program launched in Australia. This government incentive picked up old refrigerators, paying their owners for "donating" the refrigerator. The refrigerator was then refurbished, with new door seals, a thorough cleaning and the removal of items, such as the cover that is strapped to the back of many older units. The resulting refrigerators, now over 10% more efficient, were then distributed to low income families.

Effect on Lifestyle

The refrigerator allows the modern family to keep food fresh for longer than before. The most notable improvement is for meat and other highly perishable wares, which needed to be refined to gain anything resembling shelf life. (On the other hand, refrigerators and freezers can also be stocked with processed, quick-cook foods that are less healthy.) Refrigeration in transit makes it possible to enjoy foodstuffs from distant places.

Dairy products, meats, fish, poultry and vegetables can be kept refrigerated in the same space within the kitchen (although raw meat should be kept separate from other foodstuffs for reasons of hygiene).

Freezers allow people to buy food in bulk and eat it at leisure, and bulk purchases save money. Ice cream, a popular commodity of the 20th century, could previously only be obtained by traveling to where the product was made and eating it on the spot. Now it is a common food item. Ice on demand not only adds to the enjoyment of cold drinks, but is useful for first-aid, and for cold packs that can be kept frozen for picnics or in case of emergency.

Temperature Zones and Ratings

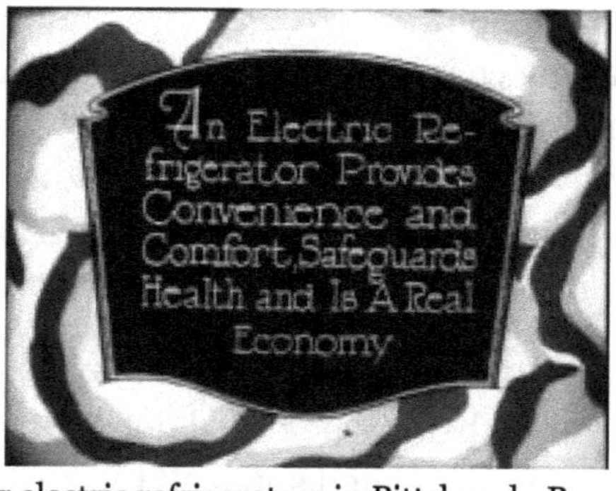

Commercial for electric refrigerators in Pittsburgh, Pennsylvania, 1926.

Some refrigerators are now divided into four zones to store different types of food:

- −18 °C (0 °F) (freezer)
- 0 °C (32 °F) (meat zone)
- 5 °C (41 °F) (cooling zone)
- 10 °C (50 °F) (crisper)

The capacity of a refrigerator is measured in either liters or cubic feet. Typically the volume of a combined refrigerator-freezer is split with 1/3rds to 1/4th of the volume allocated to the freezer although these values are highly variable.

Temperature settings for refrigerator and freezer compartments are often given arbitrary numbers by manufacturers (for example, 1 through 9, warmest to coldest), but generally 3 to 5 °C (37 to 41 °F) is ideal for the refrigerator compartment and −18 °C (0 °F) for the freezer. Some refrigerators must be within certain external temperature parameters to run properly. This can be an issue when placing units in an unfinished area, such as a garage. European freezers, and refrigerators with a freezer compartment, have a four star rating system to grade freezers.

- [?] : min temperature = −6 °C (21 °F). Maximum storage time for (pre-frozen) food is 1 week
- [??] : min temperature = −12 °C (10 °F). Maximum storage time for (pre-frozen) food is 1 month
- [???] : min temperature = −18 °C (0 °F). Maximum storage time for (pre-frozen) food is between 3 and 12 months depending on type (meat, vegetables, fish, etc.)
- [*???] : min temperature = −18 °C (0 °F). Maximum storage time for pre-frozen or frozen-from-fresh food is between 3 and 12 months

Although both the three and four star ratings specify the same storage times and same minimum temperature of −18 °C (0 °F), only a four star freezer is intended for freezing fresh food, and may include a "fast freeze" function (runs the compressor continually, down to as low as −26 °C (−15 °F)) to facilitate this. Three (or fewer) stars are used for frozen food compartments that are only suitable for storing frozen food; introducing fresh food into such a compartment is likely to result in unacceptable temperature rises. This difference in categorisation is shown in the design of the 4-star logo, where the "standard" three stars are displayed in a box using "positive" colours, denoting the same normal operation as a 3-star freezer, and the fourth star showing the additional fresh food/fast freeze function is prefixed to the box in "negative" colours or with other distinct formatting.

Most European refrigerators include a moist cold refrigerator section (which does require (automatic) defrosting at irregular intervals) and a (rarely frost free) freezer section.

Sweet Potato Storage

The sweet potato (*Ipomoea batatas*) is a very important crop for subsistence farmers in Africa and other developing countries. Due to its relatively short growing period, tolerance to drought and high yield from poor soils, it is used as a famine reserve for many of these households. However it is a highly perishable food source that is susceptible to destruction by microorganisms, metabolic spoilage, physical destruction and pests. As such they are not generally stored for extended periods of time following harvest. This is a major barrier for the optimal use of the crop and causes a lot of waste.

The majority of subsistence farmers who eat sweet potato do not utilize a storage technique of any kind, instead using a method of continuous cultivation and leaving the tubers in the ground until they are ready to be used. While less common, simple methods of storage do exist. The common ones include pit, clamp, and indoor storage and all extend the storage time of sweet potatoes by some degree. Maintaining proper storage temperature is probably the most significant barrier to good storage in subsistence farmers.

Pretreatment of sweet potato can help to minimize risk of losses. Curing can toughen the skin and heal minor physical damages while drying can reduce spoilage and inactivate metabolic degradation.

Optimum storage of sweet potatoes occurs at 12–16 °C (54–61 °F), 85–90% relative humidity (RH) and requires proper ventilation to remove excess carbon dioxide (CO_2) and bring in oxygen (O_2) for respiration. At these conditions sweet potatoes have been shown to last 5 months to a maximum of a year compared to 2–3 months normally.

Susceptibility to Spoilage

Sweet potatoes are a highly perishable crop and difficult to store for extended periods of time. This is mainly due to their high moisture content, metabolic activity following harvesting and thin, permeable skin. Following harvest, sweet potatoes are susceptible to spoilage by physical, physiological, pathological, pests and environmental means.

Physical

Physical losses refer to sweet potatoes damaged by mechanical force during harvest, handling, and transportation of the tubers. This includes injury by way of cuts, bruising, and skinning. Delicate skin and poor harvesting and handling techniques are the main cause of these issues. Damage to tubers by physical means can be minimized careful handling and storing in boxes and cartons instead of cloth sacks.

Physiological

Physiological processes within the plant can contribute highly to spoilage. Natural transpiration causes water loss and results in an unappealing pithy texture. Respiration also continues to occur after harvesting and a buildup of CO_2 around the crop can cause early spoilage. Sprouting will also occur during long term storage, which is useful for planting but will utilize nutrients in the tuber. Sprouting can be inhibited by preharvest spraying with maleic hydrazide, treatment of tubers with methyl ester of napthalene acetic acid (MENA) in acetone on paper spread between tubers and 0.5–4.0% thiourea solution.

Pathological and pests

Sweet potatoes are susceptible to a number of fungi, bacteria and viruses. Infection can cause the production of toxins, pathogenic bacteria, bitter flavours, tissue breakdown, and surface blemishes. Two common sources of infection are from *Rhizopus oryzae* (soft rot) and *Botryodiplodia* (Java black rot). These are responsible for 78% of sweet potato infections in Bangladesh. The pest of greatest concern for farmers is the sweet potato weevil (*Cylas spp.*). This can enter the tuber prior to harvesting and proliferate in storage and has been known to cause up to 50% loss in yields. Simple techniques can be used to minimize the threat of the sweet potato weevil. It has been shown that storage of the potatoes in temperatures under 20 °C (68 °F) can kill up to 89.5% of weevils in an infected crop. Dipping of tubers in hot water also been shown as an effective intervention to control infestations . Damage to tuber by physical and physiological means can increase chance of pathological infection and pests.

Environmental

The length of time that sweet potatoes may be stored is highly dependent on temperature and relative humidity. Storage in temperatures less than 10 °C (50 °F) will cause chill injury in sweet potatoes. This is characterized by discoloration of the insides of the tuber, which can range from brown to black instead of the usual orange. Formation of hard areas also occurs that remain even after cooking. Other signs of chill injury include production of off flavours, internal cell structure breakdown and increased rotting due to reduced skin formation, which allows fungal infection. On the other end of the scale, temperatures above 16 °C (61 °F) will cause a rise in the rates of respiration and sprouting in tubers as well as increased activity by microorganisms. Relative humidity under 80% will cause excessive water loss while RH above 90% will cause condensation to form, which accelerates rotting.

Ideal Storage Conditions

To minimize potential spoilage, precise storage conditions should be met. Sweet potatoes should be stored at between 12 and 16 °C (54 and 61 °F) with a RH of 85-90%. Proper ventilation of the storage area is a key factor in maximizing potential storage time. Respiration causes the production of CO2, which will act to increase the rate of spoilage. Sweet potatoes can convert approximately 57dm3 of oxygen/ton/day to carbon dioxide and need ventilation to compensate for this. The O2 in the storage environment should not fall below 7% and CO2 must not exceed 10%. Due to lack of proper equipment, maintaining adequate temperature conditions are the largest hurdle that farmers in a developing nation will have to face.

Not all sweet potatoes are suitable for storage. Tubers that have been mechanically damaged, infected with pests or pathogens, or are rotten or sprouting should not be kept. These factors will increase rate of spoilage and can increase losses in the rest of the stored crop. Under these optimum conditions sweet potatoes have been shown to keep for 5 months up to a max of about a year.

Storage Methods

In developed countries proper refrigeration technology is used to keep sweet potatoes for long periods of time. Many poor farmers do not have access to refrigeration or even electricity so must rely on other storage methods. There are several methods of storage that only require simple and cheap building materials. The most commonly used are pit storage, clamp storage and indoor storage. These techniques are used only sporadically however as most farmers practice sequential harvesting in which tubers are left in-ground until needed. Such as the Stanley Potato House.

In-ground

In subsistence farming, sweet potatoes are commonly left in the ground and eaten or sold directly following harvest, this is called piece-meal or sequential harvesting. Sweet potatoes are delicate and easily damaged. In-ground storage is used to protect the tubers while reducing the work required to set up storage facilities. In areas with reliable rainfall patterns farmers are able to maintain a supply of fresh sweet potatoes for most of the year by continuous cultivation and harvesting. In drier places however, such as northern Uganda, this is not a practical solution. This is not a long-term option and the plants will degrade after a short period of time. There are numerous disadvantages to this method as it takes up field space that could otherwise be used for growing more crops.

Pit Storage

This is a simple, cheap method that involves digging a hole in the ground for storing the potatoes. Pit storage differs from in-ground as tubers are collected and kept together and considerations are made to control the storage environment.

Construction of the pit can vary according to what materials are available but commonly used are; grass, soil, wood, lime, sawdust and ash. Grass is typically used to line the bottom and sides of the pit in order to insulate against temperature change and absorb moisture. Before usage, grass should be flamed in order to destroy any pests that may be hiding within. Soil is used to seal the roof of the pit and as filler. Wood and plant material can be used to strengthen walls as well as create a roof covering the pit. Roofs built over the pit structure are beneficial in that they can help keep rain out and provide shade to lower temperature. Lime may have some effectiveness in absorbing CO_2 and removing it form the environment. Sawdust is used as a cushioning material and to help control condensation on the tubers. Wood ash can applied to potatoes prior to storage and has shown some effectiveness in protecting against insect attacks and mould.

In order to prevent rotting the pit should ideally be dug into a slope to promote drainage. It should also be situated at least 0.35 metres (1.1 ft) above the water table during dry season and must remain above during wet/rainy season and be approximately 0.5 metres (1.6 ft) below the surface.

There can be ventilation problems with this storage method. Commonly the pit will be sealed shut in order to maintain a good RH. The draw back to this is that CO_2 will collect and cause spoilage. To help this, a sizable headspace should be maintained in the pit to help promote airflow.

Pits can be reused but they should be cleaned, the soil turned over and disinfected with fire or sulphur to rid the area of any microorganisms.

Losses are still fairly high for this technique, though it has been shown in some instances to keep roots for 2–4 months.

Clamp and Mound Storage

Another simple and low cost method, clamp storage consists of covered piles of sweet potatoes. After selecting for the tubers in the best condition, they are stacked in a heap on a layer of grass and covered in layers of grass and soil. As with pit storage ash, lime and sawdust can be used for added effect. The piles may be made at ground level or in shallow or deep trenches. Drainage should be considered and ruts may be made in the ground to lead off water. The clamps may be covered by a roof or kept in a building for added protection. To minimize losses due to respiration a ventilation shaft can be added. Results are fairly poor with this technique and estimated storage time is 2–3 months.

Indoor Storage

Sweet potatoes may be harvested and stored in building. This could include in the living area or in a granary built specifically to store produce. In home storage is typically done in straw woven baskets, cloth bags or wooden boxes. Baskets and boxes have been shown to be more effective at minimizing mechanical damage. If possible tubers should be kept off of the ground to keep them away from rodents and other pests. This is an effective technique for maintaining proper ventilation though depending on the type of building maintaining proper storage temperature and RH may be difficult.

Granaries or other storage buildings typically consist of a round hut with walls made of straw, mud, clay and wood and a conical straw roof. These are commonly supported above ground by a system of legs to keep the crop dry and away from animals, rodents and pests.

Pretreatment of Sweet Potatoes

Typically sweet potatoes are stored and eaten fresh. However there are some simple methods used to increase their storability that can be used in conjunction with other storage methods. Drying and curing of sweet potatoes are two common methods to prepare them prior to storage.

Drying of tubers is done on those that are too damaged to be stored fresh but still have edible material on them. This involves slicing tubers to a thickness of approximately 2–4 mm and then laying them out in the sun for four days or until they are rid of most of their moisture. During drying the potatoes can be covered in prickly bushes or thorns to ward off animals. Dried slices can be kept in-doors or in raised silos until eaten. Drying removes moisture, reduces bacterial growth, and inactivates metabolic processes and enzymatic decomposition.

Curing is a technique done to toughen up the outer layer of skin (periderm). It is a measure that prevents against excessive moisture loss, entry of microorganisms into the plant and even facilitates healing of mechanically damaged tubers. Curing is done by exposing the whole potatoes to a moderately high temperature for several days immediately following harvest.; 30–32 °C (86–90 °F) and 80–95% RH for 4–10 days has been found to be adequate curing conditions A common method to cure without use of incubation equipment is to wrap tubers in black polythene sheets and leave out in the sun for 5 days. The black sheet keeps moisture in and collects heat to reach necessary conditions for curing. Immediately following curing the temperature must be dropped quickly in order to avoid sprouting damage.

References

- Mitchell, Deborah (2004). Safe foods: the A-to-Z guide to the most wholesome foods for you and your family. Penguin. pp. Ch. 15. ISBN 978-1-101-21015-4

- Mitchell, Deborah (2004). Safe foods: the A-to-Z guide to the most wholesome foods for you and your family. Penguin. pp. Ch. 15. ISBN 978-1-101-21015-4
- Marotz, Lynn R. (2008). Health, Safety, and Nutrition for the Young Child. Wadsworth Publishing. pp. 491–2. ISBN 978-1-4283-2070-3
- Leslie Crutchfield and Heather McLeod Grant (2007). "Chpt 3". Forces for Good: The Six Practices of High-Impact Nonprofits. Jossey-Bass. ISBN 0470580348
- Andrew Walter (2012). William A Dando, ed. Food and Famine in the 21st Century. ABC-CLIO. pp. 171–181. ISBN 978-1-59884-730-7
- Arora, Ramesh Chandra. "Mechanical vapour compression refrigeration". Refrigeration and Air Conditioning. New Delhi, India: PHI Learning. p. 3. ISBN 81-203-3915-0
- Dennis R. Heldman (29 August 2003). Encyclopedia of Agricultural, Food, and Biological Engineering (Print). CRC Press. p. 350. ISBN 978-0-8247-0938-9

Chapter 6

Foodborne Disease Caused by Microbes

Any disease caused by food contaminated with pathogenic bacteria, viruses or parasites is known as microbial foodborne disease. Some of the significant topics studied under microbial foodborne diseases are foodborne illness, foodborne infection and intoxication, prevention of foodborne illness, treatment of foodborne illness, etc. These topics are crucial for a complete understanding of the subject.

Foodborne Illness

Foodborne illnesses are infections or irritations of the gastrointestinal (GI) tract caused by food or beverages that contain harmful bacteria, parasites, viruses, or chemicals. The GI tract is a series of hollow organs joined in a long, twisting tube from the mouth to the anus. Common symptoms of foodborne illnesses include vomiting, diarrhea, abdominal pain, fever, and chills.

Most foodborne illnesses are acute, meaning they happen suddenly and last a short time, and most people recover on their own without treatment. Rarely, foodborne illnesses may lead to more serious complications. Each year, an estimated 48 million people in the United States experience a foodborne illness. Foodborne illnesses cause about 3,000 deaths in the United States annually.

Reasons for Foodborne Illnesses

The majority of foodborne illnesses are caused by harmful bacteria and viruses.[2] Some parasites and chemicals also cause foodborne illnesses.

Bacteria

Bacteria are tiny organisms that can cause infections of the GI tract. Not all bacteria are harmful to humans.

Some harmful bacteria may already be present in foods when they are purchased. Raw foods including meat, poultry, fish and shellfish, eggs, unpasteurized milk and dairy products, and fresh produce often contain bacteria that cause foodborne illnesses. Bacteria can contaminate food—making it harmful to eat—at any time during growth, harvesting or slaughter, processing, storage, and shipping.

Foods may also be contaminated with bacteria during food preparation in a restaurant or home kitchen. If food preparers do not thoroughly wash their hands, kitchen utensils, cutting boards, and other kitchen surfaces that come into contact with raw foods, cross-contamination—the spread of bacteria from contaminated food to uncontaminated food—may occur.

If hot food is not kept hot enough or cold food is not kept cold enough, bacteria may multiply. Bacteria multiply quickly when the temperature of food is between 40 and 140 degrees. Cold food should be kept below 40 degrees and hot food should be kept above 140 degrees. Bacteria multiply more slowly when food is refrigerated, and freezing food can further slow or even stop the spread of bacteria. However, bacteria in refrigerated or frozen foods become active again when food is brought to room temperature. Thoroughly cooking food kills bacteria.

Many types of bacteria cause foodborne illnesses. Examples include

- Salmonella, a bacterium found in many foods, including raw and undercooked meat, poultry, dairy products, and seafood. Salmonella may also be present on egg shells and inside eggs.
- Campylobacter jejuni (C. jejuni), found in raw or undercooked chicken and unpasteurized milk.
- Shigella, a bacterium spread from person to person. These bacteria are present in the stools of people who are infected. If people who are infected do not wash their hands thoroughly after using the bathroom, they can contaminate food that they handle or prepare. Water contaminated with infected stools can also contaminate produce in the field.
- Escherichia coli (E. coli), which includes several different strains, only a few of which cause illness in humans. E. coli O157:H7 is the strain that causes the most severe illness. Common sources of E. coli include raw or undercooked hamburger, unpasteurized fruit juices and milk, and fresh produce.
- Listeria monocytogenes (L. monocytogenes), which has been found in raw and undercooked meats, unpasteurized milk, soft cheeses, and ready-to-eat deli meats and hot dogs.
- Vibrio, a bacterium that may contaminate fish or shellfish.
- Clostridium botulinum (C. botulinum), a bacterium that may contaminate improperly canned foods and smoked and salted fish.

Viruses

Viruses are tiny capsules, much smaller than bacteria, that contain genetic material. Viruses cause infections that can lead to sickness. People can pass viruses to each other.

Viruses are present in the stool or vomit of people who are infected. People who are infected with a virus may contaminate food and drinks, especially if they do not wash their hands thoroughly after using the bathroom.

Common sources of foodborne viruses include:

- food prepared by a person infected with a virus.
- shellfish from contaminated water.
- produce irrigated with contaminated water.

Common foodborne viruses include:

- norovirus, which causes inflammation of the stomach and intestines.
- hepatitis A, which causes inflammation of the liver.

Parasites

Parasites are tiny organisms that live inside another organism. In developed countries such as the United States, parasitic infections are relatively rare.

Cryptosporidium parvum and Giardia intestinalis are parasites that are spread through water contaminated with the stools of people or animals who are infected. Foods that come into contact with contaminated water during growth or preparation can become contaminated with these parasites. Food preparers who are infected with these parasites can also contaminate foods if they do not thoroughly wash their hands after using the bathroom and before handling food.

Trichinella spiralis is a type of roundworm parasite. People may be infected with this parasite by consuming raw or undercooked pork or wild game.

People Prone to Foodborne Illnesses

Anyone can get a foodborne illness. However, some people are more likely to develop foodborne illnesses than others, including:

- infants and children.
- pregnant women and their fetuses.
- older adults.
- people with weak immune systems.

These groups also have a greater risk of developing severe symptoms or complications of foodborne illnesses.

Symptoms of Foodborne Illnesses

Symptoms of foodborne illnesses depend on the cause. Common symptoms of many foodborne illnesses include.

- vomiting
- diarrhea or bloody diarrhea
- abdominal pain
- fever
- chills

Symptoms can range from mild to serious and can last from a few hours to several days.

C. botulinum and some chemicals affect the nervous system, causing symptoms such as:

- headache
- tingling or numbness of the skin
- blurred vision
- weakness
- dizziness
- paralysis

Diagnosis of Foodborne Illnesses

To diagnose foodborne illnesses, health care providers ask about symptoms, foods and beverages recently consumed, and medical history. Health care providers will also perform a physical examination to look for signs of illness.

Diagnostic tests for foodborne illnesses may include a stool culture, in which a sample of stool is analyzed in a laboratory to check for signs of infections or diseases. A sample of vomit or a sample of the suspected food, if available, may also be tested. A health care provider may perform additional medical tests to rule out diseases and disorders that cause symptoms similar to the symptoms of foodborne illnesses.

If symptoms of foodborne illnesses are mild and last only a short time, diagnostic tests are usually not necessary.

Eating, Diet, and Nutrition

The following steps may help relieve the symptoms of foodborne illnesses and prevent dehydration in adults;

- drinking plenty of liquids such as fruit juices, sports drinks, caffeine-free soft drinks, and broths to replace fluids and electrolytes.
- sipping small amounts of clear liquids or sucking on ice chips if vomiting is still a problem.
- gradually reintroducing food, starting with bland, easy-to-digest foods such as rice, potatoes, toast or bread, cereal, lean meat, applesauce, and bananas.
- avoiding fatty foods, sugary foods, dairy products, caffeine, and alcohol until recovery is complete.

Infants and children present special concerns. Infants and children are likely to become dehydrated more quickly from diarrhea and vomiting because of their smaller body size. The following steps may help relieve symptoms and prevent dehydration in infants and children:

- giving oral rehydration solutions such as Pedialyte, Naturalyte, Infalyte, and CeraLyte to prevent dehydration.
- giving food as soon as the child is hungry.
- giving infants breast milk or fullstrength formula, as usual, along with oral rehydration solutions.

Older adults and adults with weak immune systems should also drink oral rehydration solutions to prevent dehydration.

Traveler's Diarrhea

People who visit certain foreign countries are at risk for traveler's diarrhea, which is caused by eating food or drinking water contaminated with bacteria, viruses, or parasites. Traveler's diarrhea can be a problem for people traveling to developing countries in Africa, Asia, Latin America, and the Caribbean. Visitors to Canada, most European countries, Japan, Australia, and New Zealand do not face much risk for traveler's diarrhea.

To prevent traveler's diarrhea, people traveling from the United States to developing countries should avoid:

- drinking tap water, using tap water to brush their teeth, or using ice made from tap water.
- drinking unpasteurized milk or milk products.
- eating raw fruits and vegetables, including lettuce and fruit salads, unless they peel the fruits or vegetables themselves.
- eating raw or rare meat and fish.

- eating meat or shellfish that is not hot when served
- eating food from street vendors

Travelers can drink bottled water, bottled soft drinks, and hot drinks such as coffee or tea.

People concerned about traveler's diarrhea should talk with a health care provider before traveling. The health care provider may recommend that travelers bring medication with them in case they develop diarrhea during their trip. Health care providers may advise some people—especially people with weakened immune systems—to take antibiotics before and during a trip to help prevent traveler's diarrhea. Early treatment with antibiotics can shorten a bout of traveler's diarrhea.

Types of Foodborne Illness

There are dozens of different bacteria, viruses, parasites and other pathogens which can contaminate raw or improperly prepared food thereby leading to human infection and the transmission of various foodborne diseases. However, the CDC estimates that about 90% of all foodborne illness in this country is caused by the following seven (7) pathogens: Norovirus, Salmonella, Clostridium perfrigens, Campylobacter, Listeria, E. coli 0157:H7 and Toxoplasma. While most people recover from food poisoning within a few days, every year nearly 130,000 Americans suffer a serious foodborne illnesses which requires hospitalization. In some cases, the foodborne disease proves fatal or leads to long term health consequences. The following are the most significant foodborne illnesses based upon frequency of occurrence and/or severity of potential adverse health consequences:

- Salmonellosis - a potentially life threatening infectious disease caused by various species of Salmonella bacteria. This infection can cause a debilitating condition referred to as Reactive Arthritis ("Reiter's Syndrome").
- Hemorrhagic Colitis - infectious disease primarily caused by the bacterium *Escherichia coli* O157:H7. E. coli infection can lead to a serious complication known as Hemolytic Uremic Syndrome (HUS).
- Listeriosis - infectious disease caused by the bacterium *Listeria monocytogenes*. This infection is particularly dangerous for pregnant women because it can cause stillbirth, premature labor and/or infection of the baby.
- Campylobacteriosis - infectious disease caused by the bacterium *Campylobacter jejuni*. This infection can lead to a serious disorder affecting the peripheral nervous system known as Guillain – Barre Syndrome.
- Shigellosis (bacillary dysentery) - infectious disease caused by various species of Shigella bacteria. This infection can also cause HUS or Reactive Arthritis.

- Perfringens Poisoning - enteric disease caused by the bacterium *Colostridium perfrigens.*
- Foodborne Botulism - life threatening disease caused by consumption of foods containing neurotoxins produced by the bacterium *Colostridium botulinum.*
- Norovirus Gastroenteritis - highly contagious disease caused by the Norwalk virus family. This viral infection can sometimes result in a chronic functional gastrointestinal disorder known as Post Infectious Irritable Bowel Syndrome.
- Acute Viral Hepatitis A - infectious disease of the liver caused by the Hepatitis A virus (HAV).
- Toxoplasmosis - infectious disease caused by a microscopic parasite called *Toxoplasma gondii.* In pregnant women, this infection can be transmitted and harmful to the fetus.

It is estimated that more than half of all foodborne illnesses are caused by some type of viral pathogen, while about one-third are caused by various species of bacteria. Parasites are responsible for a relatively small percentage (<5%) of foodborne illnesses in this country. Hepatitis A and Norovirus are highly contagious infectious diseases caused by viral pathogens. E. coli food poisoning, Salmonellis, Listeriosis, Campylobacteriosis and Shigellosis are communicable infectious diseases caused by different bacterial pathogens. Toxins produced by different species of bacteria in the Clostridium family cause Botulism and Perfringens poisoning which, unlike the other aforementioned illnesses, are not spread via person-to-person transmission from an infected individual to another previously non-infected individual. Toxoplasmosis is an infectious disease caused by a parasite. All of the pathogens which cause these illnesses – except Listeria, Hepatitis A and Toxoplasma – first attack the intestines causing acute gastrointestinal illness which may or may not lead to other illnesses or complications. Listeria, Hepatitis A and Toxoplasma infections generally cause various types of non-gastrointestinal illnesses. Listeria and Toxoplasma infections are unique in that they can be transmitted by an infected pregnant woman to her unborn child.

Some types of foodborne illness caused by bacteria can be treated with antibiotics with varying effectiveness. However, in certain types of bacterial food poisoning, antibiotics may be contraindicated. For example, the CDC recommends that E. coli infections not be treated with antibiotics because of a concern that they may increase the risk of Hemolytic Uremic Syndrome. Antibiotics are also not recommended for treatment of Perfringens poisoning. There is no “cure” for the viral infections, and they cannot be treated with antibiotics. However, acute viral Hepatitis A can be prevented with proper immunization. Unfortunately, Hepatitis A is the only one of these foodborne illnesses for which there currently is a vaccine.

The Most Dangerous Pathogens Causing Foodborne Illness

The most often occurring foodborne illnesses are Noroviruses, Salmonellosis, Perfringens poisoning and Campylobacteriosis, in that order. Fortunately, all of these foodborne illnesses, except Salmonellosis, have a very low mortality rate and incidence of serious complications; Salmonellosis results in significant morbidity. Indeed, Salmonella is the leading cause of death among all known foodborne pathogens. In addition to Salmonella, the foodborne pathogens of greatest concern include Toxoplasma gondii, Listeria monocytogenes, Norovirus, Campylobacter jejuni and E. coli (STEC) 0157. According to the CDC, these six pathogens are to blame for more than 90% of all food poisoning deaths in the United States. Moreover, E. coli O157:H7 and Salmonella deserve particular attention because these pathogens are responsible for many of the most notorious food poisoning outbreaks which have taken place in the last two decades. Listeria and Toxoplasma pathogens are especially concerning for pregnant women because of their ability to infect both mother and fetus, and cause miscarriages, stillbirths or other complications during pregnancy.

The Potential Complications of Foodborne Illnesses and Persons at Highest Risk of Suffering Them

Fortunately, the majority of food poisoning episodes resolve within a week and without significant adverse health consequences. However, the risk of death, hospitalization and debilitating medical conditions is present with any foodborne illness, especially among certain groups of people considered most at risk. The categories of persons known to be at higher risk of contracting a foodborne illness and/or suffering death or serious medical complications therefrom include the following:

- Infants and young children.
- Pregnant women (and their unborn fetus).
- Elderly adults over the age of 65.
- Persons with a weakened or compromised immune system.
- Cancer patients especially those undergoing chemotherapy or radiation treatments.
- Persons who are HIV positive or suffering AIDS.
- Persons with a certain chronic conditions such as diabetes, liver disease or kidney failure.
- Persons who have sickle cell anemia or other disorders affecting red blood cells.

- Transplant recipients especially while taking medications to reduce the risk of rejection.
- Persons taking antibiotics, antihistamines or steroid medications.

Food poisoning can be fatal, although the mortality rate in this country is very low. Death occurs in less than 1 out of every 10,000 cases of food poisoning. However, even with this low incident rate, the number of deaths is significant because the number of food poisoning episodes is so high (about 48 million cases are believed to occur annually). The U.S. food poisoning death toll is currently estimated by the CDC to be 3,000 per year, although some experts believe the annual number of deaths due to foodborne illness could be significantly higher. According to the USDA Economic Research Service, "foodborne illnesses account for about 1 of every 100 U.S. hospitalizations and for 1 of every 500 U.S. deaths." In cases of foodborne illness, the causes of death are varied but frequently include septicemia (which occurs after a bacterial infection has invaded the blood stream) and/or failure of one or more organs due to metastatic infection or some other disease process. For pregnant women, their unborn child is also at risk. Some foodborne illnesses (primarily Listeriosis and Toxoplasmosis) can cause spontaneous abortion and stillbirth of the fetus.

The potential health implications of a nonfatal foodborne illness are often under appreciated. According to the latest estimates released by the CDC in 2011, nearly 130,000 food poisoning victims require hospitalization in this country every year. Many of these patients face a wide array of potential complications with some having profound long term health consequences. Depending upon the type of foodborne illness involved, the list of known food poisoning complications includes diabetes, kidney dysfunction, urinary tract infections, renal induced hypertension, pernicious anaemia, hemorrhagic adrenalitis, partial paralysis, meningitis, encephalitis, brain damage, cognitive impairment, autoimmune disorders, glaucoma, vision impairment, endocarditis, pericarditis, lactose intolerance, gastroesophageal disease (GERD), dyspepsia, gastroparesis, and pseudomemdraneous colitis.

The Four Most Prevalent Food Poisoning Complications

Of particular note, the four most concerning and frequently cited medical conditions resulting from foodborne illnesses are as follows:

- Hemolytic Uremic Syndrome - a serious disease that occurs when an enteric bacterial infection leads to the production of toxins that enter the bloodstream and cause destruction of red blood cells. This can lead to further complications including hemolytic anemia, thrombocytopenia (blood platelet deficiency), uremia, pancreatitis, neurological dysfunction, seizures, acute respiratory distress syndrome (ARDS), renal damage and acute kidney failure. Children are at especially high risk for HUS. Most persons suffering HUS require hospitalization of signification duration. Most eventually recover but about one-third of patients suffer permanent sequellae, and the U.S. fatality rate is 3-5%. Most cases of HUS are associated with E. coli food poisoning, but this complication can also occur in cases of Shigelliosis.

- Guillain-Barre Syndrome (GBS) - a serious disorder of the peripheral nervous system caused by the body's autoimmune response to the presence of foreign pathogens. Among other things, GBS can lead to varying degrees of neuromuscular paralysis. Campylobacter is the primary cause of GBS.
- Reactive Arthritis (also called "Reiter's Syndrome") - a potentially debilitating disorder resulting from the body's autoimmune release of antigens to attack bacteria which have entered the bloodstream following a gastrointestinal infection. This process can cause joint inflammation and various other arthritic like conditions. This complication is known to occur in cases of food poisoning caused by Salmonella, Campylobacter, E. coli and Shigella.
- Post-Infectious Irritable Bowel Syndrome (IBS) - a potentially chronic and long-term functional gastrointestinal disorder characterized by various symptoms (which can occur daily or at irregular intervals) such as bloating, abdominal pain, diarrhea and constipation. This condition can be caused or exacerbated by severe gastrointestinal infections which occur after food poisoning. Foodborne illnesses caused by E. coli 0157:H7, Salmonella, Shigella, Campylobacter and Norovirus are known to be associated with IBS.

Foodborne Infection and Intoxication

Foodborne Infection

A foodborne infection is caused by the ingestion of food containing pathogenic microorganisms (i.e bacteria, virus or parasite) which must multiply with in the gastrointestinal tract, producing widespread inflammation. The most commonly implicated microorganisms include species of *Salmonella, Shigella, E. coli 0157:H7,* etc. These infections have longer incubation periods than those experienced with food intoxications, usually commencing from 6-24 hours or longer after ingestion. Symptoms may include fever, headache, nausea, vomiting, diarrhea, abdominal pain or distress, and prostration. The causative organism may be identified by laboratory examination of the vomitus, feces, or blood and the suspected food, when available.

Foods most commonly incriminated in outbreaks of foodborne infections are meat and seafood mixtures such as hash, hamburger, creamed meat pies, crab, lobster, chicken, and turkey salads, turkey, turkey stuffing or dressing, chicken, and ham. These foods have common characteristics in that they provide moisture, a good protein food supply and warmth. Given sufficient time, these factors promote an ideal environment for the growth and multiplication of microorganisms. It is important to remember that these organisms do not necessarily cause any alteration in the normal appearance, odor, or taste of the food.

Foodborne Intoxication

Certain bacteria under favorable growth conditions produce chemicals (toxins) in food which when ingested will cause food intoxication. Enterotoxins produced by *Staphylococcus aureus* are heat stable (i.e., not destroyed by normal cooking temperatures) and are the cause of the most common foodborne intoxication. The staphylococci multiply in the food where they produce their toxins before the food is consumed. It generally takes less than 8 hours for these organisms to elaborate enough toxins to cause symptoms. The disease is characterized by an abrupt onset (2 to 4 hours after ingestion) of symptoms of severe nausea, vomiting, diarrhea, and prostration with little or no fever.

Staphylococcal food intoxication usually follows ingestion of starchy food, especially potato salad, custard and pies. When the offending food is meat, pork (including ham and salami) and poultry products are usually the source. Ham may become contaminated with staphylococci during the practice of boning, slicing and holding without adequate refrigeration for several hours before serving. In addition, highly salted ham permits staphylococcal growth but inhibits many other bacteria. Other foods commonly involved are canned or potted meat or fish, pressed tongue, beef, cheese, other milk products, cream or custard filled pastries, potato salad, and pasta salads. The usual source of the pathogens, which cause this form of food intoxication, may be the nose, throat, boils, pimples, or infected cuts on the hands of food service personnel.

Prevention of Foodborne Illness

Following are the tips useful in preventing foodborne illnesses:

- Pick up your packaged and canned foods first.
- Don't buy food in cans that are bulging or dented or in jars that are cracked or have loose or bulging lids.
- Don't eat raw shellfish and use only pasteurized milk and cheese and pasteurized or otherwise treated ciders and juices if you have a health problem, especially one that may have impaired your immune system.
- Choose eggs that are refrigerated in the store. Before putting them in your cart, open the carton and make sure that the eggs are clean and none are cracked.
- Select frozen foods and perishables such as meat, poultry or fish last. Always

put these products in separate plastic bags so that drippings don't contaminate other foods in your shopping cart.

- Don't buy frozen seafood if the packages are open, torn or crushed on the edges. Avoid packages that are above the frost line in the store's freezer. If the package cover is transparent, look for signs of frost or ice crystals. This could mean that the fish has either been stored for a long time or thawed and refrozen.
- Check for cleanliness at the meat or fish counter and the salad bar. For instance, cooked shrimp lying on the same bed of ice as raw fish could become contaminated.
- When shopping for shellfish, buy from markets that get their supplies from state-approved sources; stay clear of vendors who sell shellfish from roadside stands or the back of a truck. And if you're planning to harvest your own shellfish, heed posted warnings about the safety of the water.
- Take an ice chest along to keep frozen and perishable foods cold if it will take more than an hour to get your groceries home.

Treatment of Foodborne Illness

Treatment for food poisoning typically depends on the source of the illness, if known, and the severity of your symptoms. For most people, the illness resolves without treatment within a few days, though some types of food poisoning may last longer.

Treatment of food poisoning may include:

- Replacement of lost fluids: Fluids and electrolytes — minerals such as sodium, potassium and calcium that maintain the balance of fluids in your body — lost to persistent diarrhea need to be replaced. Some children and adults with persistent diarrhea or vomiting may need hospitalization, where they can receive salts and fluids through a vein (intravenously), to prevent or treat dehydration.
- Antibiotics: Your doctor may prescribe antibiotics if you have certain kinds of bacterial food poisoning and your symptoms are severe. Food poisoning caused by listeria needs to be treated with intravenous antibiotics during hospitalization. The sooner treatment begins, the better. During pregnancy, prompt antibiotic treatment may help keep the infection from affecting the baby.

  Antibiotics will not help food poisoning caused by viruses. Antibiotics may actually worsen symptoms in certain kinds of viral or bacterial food poisoning. Talk to your doctor about your options.

Adults with diarrhea that isn't bloody and who have no fever may get relief from taking the medication loperamide (Imodium A-D) or bismuth subsalicylate (Pepto-Bismol). Ask your doctor about these options.

Lifestyle and Home Remedies

Food poisoning often improves without treatment within 48 hours. To help keep yourself more comfortable and prevent dehydration while you recover, try the following:

- Let your stomach settle: Stop eating and drinking for a few hours.
- Try sucking on ice chips or taking small sips of water: You might also try drinking clear soda, clear broth or noncaffeinated sports drinks, such as Gatorade. You're getting enough fluid when you're urinating normally and your urine is clear and not dark.
- Ease back into eating: Gradually begin to eat bland, low-fat, easy-to-digest foods, such as soda crackers, toast, gelatin, bananas and rice. Stop eating if your nausea returns.
- Avoid certain foods and substances until you're feeling better: These include dairy products, caffeine, alcohol, nicotine, and fatty or highly seasoned foods.
- Rest: The illness and dehydration can weaken and tire you.

Bacterial Pathogens

While most species of bacteria are harmless, and others beneficial for us, there are a number of disease-causing bacteria, which are called "pathogenic bacteria." Pathogenic bacteria can contribute to many worldwide diseases, including tuberculosis, cholera, anthrax, leprosy, the bubonic plague, pneumonia, and food-borne illnesses. The most common fatal bacterial infectious diseases are respiratory infections, with tuberculosis (caused by the bacteria *Mycobacterium tuberculosis*) killing approximately two million people each year, generally in sub-Saharan Africa. Bacterial infections can be caused by a wide range of bacteria which can lead to mild illness to life-threatening illnesses (like bacterial meningitis), which require immediate hospital interventions. Infections caused by bacteria are one of the leading causes of death in both the very young and the very old. Under normal circumstances, people are protected against bacterial infections by a healthy immune system.

Pathogenic bacteria use different strategies for infection: they produce toxins, the use parasites for spreading, or only by different adhesion strategies they are causing inflammation in the body.

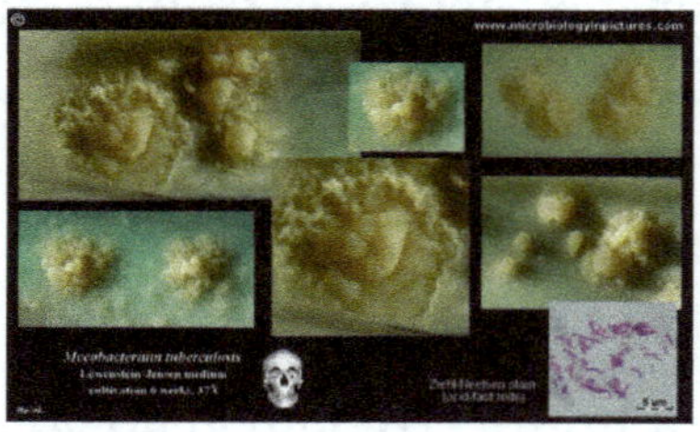

Mycobacterium tuberculosis the causative agent of tuberculosis.

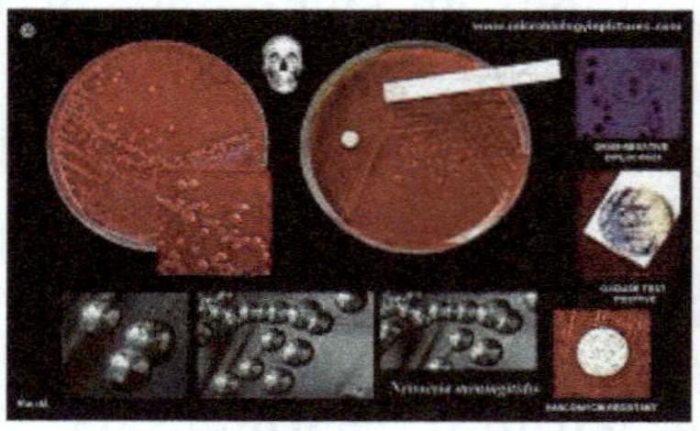

Neisseria meningitidis the causative agent of bacterial meningitis.

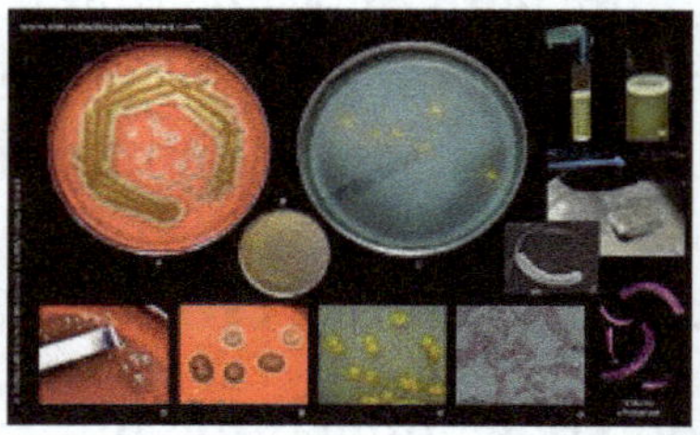

Vibrio cholerae the causative agent of cholera.

Mechanisms of Damage

The symptoms of disease appear as pathogenic bacteria damage host tissues or interfere with their function. The bacteria can damage host cells directly. They can also cause damage indirectly by provoking an immune response that inadvertently damages host cells.

Direct

Once pathogens attach to host cells, they can cause direct damage as the pathogens use the host cell for nutrients and produce waste products. For example, *Streptococcus mutans*, a component of dental plaque, metabolizes dietary sugar and produces acid as a waste product. The acid decalcifies the tooth surface to cause dental caries. However, toxins produced by bacteria cause most of the direct damage to host cells.

Toxin Production

Endotoxins are the lipid portions of lipopolysaccharides that are part of the outer membrane of the cell wall of gram negative bacteria. Endotoxins are released when the bacteria lyses, which is why after antibiotic treatment, symptoms can worsen at first as the bacteria are killed and they release their endotoxins. Exotoxins are secreted into the surrounding medium or released when the bacteria die and the cell wall breaks apart.

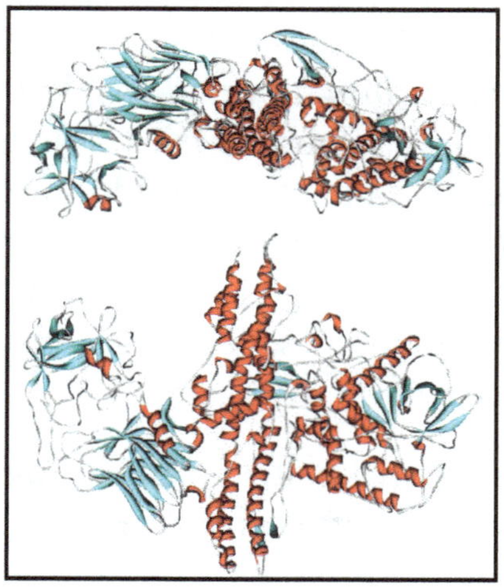

Protein structure of Botulinum toxin 3BTA.

Indirect

An excessive or inappropriate immune response triggered by an infection may damage host cells.

Survival in Host

Nutrients

Iron is required for humans, as well as the growth of most bacteria. To obtain free iron, some pathogens secrete proteins called siderophores, which take the iron away from iron-transport proteins by binding to the iron even more tightly. Once the iron-siderophore complex is formed, it is taken up by siderophore receptors on the bacterial surface and then that iron is brought into the bacterium.

Identification

Typically identification is done by growing the organism in a wide range of cultures which can take up to 48 hours. The growth is then visually or genomically identified. The cultured organism is then subjected to various assays to observe reactions to help further identify species and strain.

Treatment

Bacterial infections may be treated with antibiotics, which are classified as bacteriocidal if they kill bacteria or bacteriostatic if they just prevent bacterial growth. There are many types of antibiotics and each class inhibits a process that is different in the pathogen from that found in the host. For example, the antibiotics chloramphenicol and tetracyclin inhibit the bacterial ribosome but not the structurally different eukaryotic ribosome, so they exhibit selective toxicity. Antibiotics are used both in treating human disease and in intensive farming to promote animal growth. Both uses may be contributing to the rapid development of antibiotic resistance in bacterial populations. Phage therapy can also be used to treat certain bacterial infections.

Prevention

Infections can be prevented by antiseptic measures such as sterilizing the skin prior to piercing it with the needle of a syringe and by proper care of indwelling catheters. Surgical and dental instruments are also sterilized to prevent infection by bacteria. Disinfectants such as bleach are used to kill bacteria or other pathogens on surfaces to prevent contamination and further reduce the risk of infection. Bacteria in food are killed by cooking to temperatures above 73 °C (163 °F).

Campylobacter Jejuni

Campylobacter jejuni: A bacterium that typically infects the bowels. Now the leading cause of bacterial food poisoning, Campylobacter jejuni is most often spread by contact with raw or undercooked poultry. A single drop of juice from a contaminated chicken is enough to make someone sick. Symptoms tend to start 2 to 5 days after exposure and typically last a week. They resemble the symptoms of viral gastroenteritis'diarrhea, fever, abdominal pain, cramping, nausea, and vomiting'but with campylobacter, fever is typical and the diarrhea is often bloody. Antibiotics can be helpful treatment. Most people recover completely. However, some suffer long-term consequences, such as arthritis or Guillain-Barr' syndrome. Both are thought to occur when a person's immune system is activated by the Campylobacter jejuni and misdirected to attack the person's own body.

Disease

Campylobacteriosis is an infectious disease caused by bacteria of the genus *Campylobacter*. In most people who become ill with campylobacteriosis, symptoms develop within two to five days of exposure to the organism and illness typically lasts seven days following onset. Infection with *C. jejuni* usually results in enteritis, which is characterised by abdominal pain, diarrhea, fever, and malaise. Diarrhea itself can vary in severity from loose to bloody stools. The disease is usually self-limiting. However, it does respond to antibiotics. Severe (accompanying fevers, blood in stools) or prolonged cases may require erythromycin, azithromycin, ciprofloxacin, or norfloxacin. Fluid replacement via Oral Rehydration Salts may be needed and intravenous fluid may be required for serious cases.

Pathogenesis

Studies on the pathogenesis of *C. jejuni* show that for this organism to cause disease, the susceptibility of the host and the relative virulence of the infecting strain are both important. Infection results from the ingestion of contaminated food or water, and the infective dose can be as low as 800 organisms. To initiate infection, the organism must penetrate the gastrointestinal mucus, which it does using its high motility and spiral shape. The bacteria must then adhere to the gut enterocytes and can then induce diarrhea by toxin release. *C. jejuni* releases several different toxins, mainly enterotoxin and cytotoxins, which vary from strain to strain and correlate with the severity of the enteritis. During infection, levels of all immunoglobulin classes rise. Of these, IgA is the most important because it can cross the gut wall. IgA immobilises organisms, causing them to aggregate and activate complement, and also gives short-term immunity against the infecting strain of organism. The bacteria colonize the small and large intestines, causing inflammatory diarrhea with fever. Stools contain leukocytes and blood. The role of toxins in pathogenesis is unclear. *C jejuni* antigens that cross-react with one or more neural structures may be responsible for triggering the Guillain–Barré syndrome.

Hypoacylated lipopolysaccharide (LPS) from *C. jejuni* induces moderate TLR4-mediated inflammatory response in macrophages and such LPS bioactivity may eventually result in the failure of local and systemic bacterial clearance in patients. At the same time, moderation of anti-bacterial responses may be advantageous for infected patients in clinical practice, since such an attenuated LPS may not be able to induce severe sepsis in susceptible individuals.

Sources

C. jejuni is commonly associated with poultry, and it naturally colonises the digestive tract of many bird species. All types of poultry and wild birds can become colonized with Campylobacter. One study found that 30% of European starlings in farm settings in Oxfordshire, United Kingdom, were carriers of *C. jejuni*. It is also common in cattle, and although it is normally a harmless commensal of the gastrointestinal tract in these animals, it can cause campylobacteriosis in calves. It has also been isolated from wombat and kangaroo feces, being a cause of bushwalkers' diarrhea. Contaminated drinking water and unpasteurized milk provide an efficient means for distribution. Contaminated food is a major source of isolated infections, with incorrectly prepared meat and poultry as the primary source of the bacteria. Moreover, surveys show that 20 to 100% of retail chickens are contaminated. This is not overly surprising, since many healthy chickens carry these bacteria in their intestinal tracts. Raw milk is also a source of infections. The bacteria are often carried by healthy cattle and by flies on farms. Unchlorinated water may also be a source of infections. However, properly cooking chicken, pasteurizing milk, and chlorinating drinking water kill the bacteria. *Campylobacter* is not, in contrast to *Salmonella*, transmitted vertically and therefore humans do not get infected by consuming eggs.

Possible Complications

Local complications of *Campylobacter* infections occur as a result of direct spread from the gastrointestinal tract and can include cholecystitis, pancreatitis, peritonitis, and massive gastrointestinal hemorrhage. Extraintestinal manifestations of *Campylobacter* infection are quite rare and may include meningitis, endocarditis, septic arthritis, osteomyelitis, and neonatal sepsis. Bacteremia is detected in <1% of patients with *Campylobacter* enteritis and is most likely to occur in patients who are immunocompromised or among the very young or very old. Transient bacteremia in immunocompetent hosts with *C. jejuni* enteritis may be more common, but not detected because most strains are rapidly cleared by the killing action of normal human serum and because blood cultures are not routinely performed for patients with acute gastrointestinal illness.

Serious systemic illness caused by *Campylobacter* infection rarely occurs, but can lead to sepsis and death. The case-fatality rate for *Campylobacter* infection is 0.05 per 1000 infections. For instance, one major possible complication that *C. jejuni* can cause is Guillain–Barré syndrome, which induces neuromuscular paralysis in a sizeable percentage of those who suffer from it. Over time, the paralysis is typically reversible to some extent; nonetheless, about 20% of patients with GBS are left disabled, and around 5% die. Another chronic condition that may be associated with *Campylobacter* infection is what was formerly known as Reiter's syndrome, a form of reactive arthritis. Reactive arthritis is a complication strongly associated with a particular genetic makeup. That is, persons who have the human lymphocyte antigen B27 (HLA-B27) are most susceptible. Most often, the symptoms of reactive arthritis will occur up to several weeks after infection.

Epidemiology

Frequency

- United States

An estimated 2 million cases of *Campylobacter* enteritis occur annually, accounting for 5–7% of cases of gastroenteritis. *Campylobacter* organisms have a large animal reservoir, with up to 100% of poultry, including chickens, turkeys, and waterfowl, having asymptomatic intestinal infections. The major reservoirs of *C. fetus* are cattle and sheep. Nonetheless, the incidence of *Campylobacter* infections has been declining. Changes in the incidence of culture-confirmed *Campylobacter* infections have been monitored by the Foodborne Diseases Active Surveillance Network (FoodNet) since 1996. In 2010, *Campylobacter* incidence showed a 27% decrease compared with 1996–1998. In 2010, the incidence was 13.6 cases per 100,000 population, and this did not change significantly compared with 2006–2008.

- International

C. jejuni infections are extremely common worldwide, although exact figures are not available. New Zealand reported the highest national campylobacteriosis rate, which peaked in May 2006 at 400 per 100,000 population.

- Sex

Campylobacter organisms are isolated more frequently from males than females. Homosexual men appear to be at increased risk for infection with atypical *Campylobacter* species such as *Helicobacter cinaedi* and *Helicobacter fennelliae.*

- Age

Campylobacter infections can occur in all age groups. Studies show a peak incidence in children younger than 1 year and in persons aged 15–29 years. The age-specific attack rate is highest in young children. In the United States, the highest incidence of *Campylobacter* infection in 2010 was in children younger than 5 years and was 24.4 cases per 100,000 population, However, the rate of fecal cultures positive for *Campylobacter* species is greatest in adults and older children.

Genome

The genome of *C. jejuni* strain NCTC11168 was published in 2000, revealing 1,641,481 base pairs (30.6% G+C) predicted to encode 1,654 proteins and 54 stable RNA species. The genome is unusual in that virtually no insertion sequences or phage-associated sequences and very few repeat sequences are found. One of the most striking findings in the genome was the presence of hypervariable sequences. These short homopolymeric runs of nucleotides were commonly found in genes encoding the biosynthesis or modification of surface structures, or in closely linked genes of unknown function. The apparently high rate of variation of these homopolymeric tracts may be important in the survival strategy of *C. jejuni*.

Initial transposon mutagenesis screens revealed 195 essential genes, although this number is likely to go up with additional analysis.

Natural Genetic Transformation

C. jejuni is naturally competent for genetic transformation. Natural genetic transformation is a sexual process involving DNA transfer from one bacterium to another through the intervening medium, and the integration of the donor sequence into the recipient genome by homologous recombination. *C. jejuni* freely takes up foreign DNA harboring genetic information responsible for antibiotic resistance. Antibiotic resistance genes are more frequently transferred in biofilms than between planktonic cells (single cells that float in liquid media).

Laboratory Characteristics

Characteristic	Result
Growth at 25 °C	–
Growth at 35–37 °C	+
Growth at 42 °C	+
Nitrate reduction	+
Catalase test	+
Oxidase test	+
Growth on MacConkey agar	+
Motility (wet mount)	+
Glucose utilization	–
Hippurate hydrolysis	+
Resistance to nalidixic acid	–
Resistance to cephalothin	+

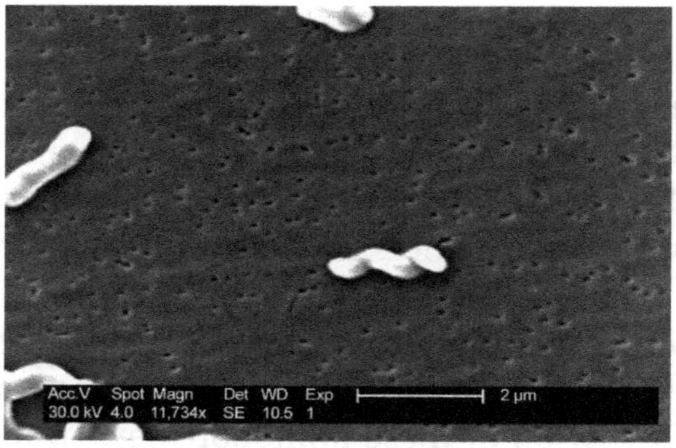

Scanning electron micrograph depicting a number of *Campylobacter jejuni* bacteria.

Under light microscopy, *C. jejuni* has a characteristic "sea-gull" shape as a consequence of its helical form. *Campylobacter* is grown on specially selective "CAMP" agar plates at 42 °C, the normal avian body temperature, rather than at 37 °C, the temperature at which most other pathogenic bacteria are grown. Since the colonies are oxidase positive, they usually only grow in scanty amounts on the plates. Microaerophilic conditions are required for luxurious growth. A selective blood agar medium (Skirrow's medium) can be used. Greater selectivity can be gained with an infusion of a cocktail of antibiotics: vancomycin, polymixin-B, trimethoprim, and actidione ([Preston's agar]), and growth under microaerophilic conditions at 42 °C.

Symptoms

Some people who have a Campylobacter infection have mild symptoms, such as a few loose stools per day. These people probably would not seek medical attention, since the symptoms would resolve on their own.

When full-blown symptoms occur, they usually begin two to seven days after exposure to Campylobacter. At first, you may have a 12- to 48-hour period of fever, headache, muscle aches, and malaise (a general feeling of sickness). These early symptoms are followed by crampy abdominal pain and diarrhea, sometimes with nausea and vomiting. There may be up to 10 loose stools watery stools per day. You may notice some blood in the stool.

Diagnosis

Your doctor may suspect that you have some sort of food-related illness based on your history and symptoms. If your symptoms are very typical and not severe, the treatment for most food-related illnesses is the same. The only way to confirm that you definitely have been infected with *Campylobacter* is to collect a stool sample for examination in a laboratory (called a stool culture). However, a stool culture is often not needed.

Some patients have symptoms that are not typical. In those cases, the doctor would send the stool culture and also recommend additional tests such as CT scan.

Expected Duration

In adults who have normal immune defenses, *Campylobacter* infection is usually a self-limited illness that goes away on its own within seven to 10 days. With antibiotic treatment, symptoms often stop sooner, usually in about five days.

Prevention

Uncooked commercial chicken, raw veal and raw ground beef are frequent sources of Campylobacter. To help prevent infection:

Cook poultry and meat thoroughly.

Wash your hands after handling raw poultry and meat.

Wash kitchen countertops and utensils thoroughly after they have been used to prepare poultry and meat.

You also can reduce your risk by never drinking unpasteurized milk or untreated water.

If your pet becomes sick with diarrhea, wash your hands thoroughly after you have cared for it. Keep sick pets away from infants, elderly family members, and family members with weakened immune systems.

Bacillus Cereus

Bacillus cereus is a Gram-positive, rod-shaped, aerobic, facultatively anaerobic, motile, beta hemolytic bacterium commonly found in soil and food. Some strains are harmful to humans and cause foodborne illness, while other strains can be beneficial as probiotics for animals. It is the cause of "fried rice syndrome", as the bacteria are classically contracted from fried rice dishes that have been sitting at room temperature for hours. *B. cereus* bacteria are facultative anaerobes, and like other members of the genus *Bacillus*, can produce protective endospores. Its virulence factors include cereolysin and phospholipase C.

The *Bacillus cereus* group comprises seven closely related species: *B. cereus sensu stricto* (referred to herein as *B. cereus*), *B. anthracis*, *B. thuringiensis*, *B. mycoides*, *B. pseudomycoides*, *B. weihenstephanensis*, and *B. cytotoxicus*.

Ecology

B. cereus competes with other microorganisms such as *Salmonella* and *Campylobacter* in the gut, so its presence reduces the numbers of those microorganisms. In food animals such as chickens, rabbits and pigs, some harmless strains of *B. cereus* are used as a probiotic feed additive to reduce *Salmonella* in the intestines and cecum. This improves the animals' growth, as well as food safety for humans who eat them. They parasitize codling moth larvae.

B. cereus and other members of *Bacillus* are not easily killed by alcohol; in fact, they have been known to colonize distilled liquors and alcohol-soaked swabs and pads in numbers sufficient to cause infection.

Some strains of *B. cereus* produce cereins, bacteriocins active against different *B. cereus* strains or other Gram-positive bacteria.

Reproduction

At 30 °C (86 °F), a population of *B. cereus* can double in as little as 20 minutes or as long as 3 hours, depending on the food product.

Food	Minutes to double, 30 °C (86 °F)	Hours to multiply by 1,000,000
Milk	20–36	6.6 - 12
Cooked rice	26–31	8.6 - 10.3
Infant formula	56	18.6

Pathogenesis

B. cereus is responsible for a minority of foodborne illnesses (2–5%), causing severe nausea, vomiting, and diarrhea. *Bacillus* foodborne illnesses occur due to survival of the bacterial endospores when food is improperly cooked. Cooking temperatures less than or equal to 100 °C (212 °F) allow some *B. cereus* spores to survive. This problem is compounded when food is then improperly refrigerated, allowing the endospores to germinate. Cooked foods not meant for either immediate consumption or rapid cooling and refrigeration should be kept at temperatures below 10 °C (50 °F) or above 50 °C (122 °F). Germination and growth generally occur between 10 °C and 50 °C, though some strains are psychrotrophic. Bacterial growth results in production of enterotoxins, one of which is highly resistant to heat and acids (pH levels between 2 and 11); ingestion leads to two types of illness, diarrheal and emetic (vomiting) syndrome.

- The diarrheal type is associated with a wide range of foods, has an 8.0- to 16-hour incubation time, and is associated with diarrhea and gastrointestinal pain. Also known as the 'long-incubation' form of *B. cereus* food poisoning, it might be difficult to differentiate from poisoning caused by *Clostridium perfringens*. Enterotoxin can be inactivated after heating at 56 °C (133 °F) for 5 minutes however it is unclear whether its presence in food causes the symptom since it degrades in stomach enzymes; its subsequent production by surviving *B. cereus* spores within the small intestine may be the cause of illness.
- The 'emetic' form is commonly caused by rice cooked for a time and temperature insufficient to kill any spores present, then improperly refrigerated. It can produce a toxin, cereulide, which is not inactivated by later reheating. This form leads to

nausea and vomiting one to five hours after consumption. It can be difficult to distinguish from other short-term bacterial foodborne intoxications such as by *Staphylococcus aureus*. Emetic toxin can withstand 121 °C (250 °F) for 90 minutes.

The diarrhetic syndromes observed in patients are thought to stem from the three toxins: hemolysin BL (Hbl), nonhemolytic enterotoxin (Nhe) and cytotoxin K (CytK). The *nhe/hbl/cytK* genes are located on the chromosome of the bacteria. Transcription of these genes is controlled by *PlcR*. These genes occur in the taxonomically related *B. thuringiensis* and *B. anthracis*, as well. These enterotoxins are all produced in the small intestine of the host, thus thwarting digestion by host endogenous enzymes. The Hbl and Nhe toxins are pore-forming toxins closely related to ClyA of *E. coli*. The proteins exhibit a conformation known as "beta-barrel" that can insert into cellular membranes due to a hydrophobic exterior, thus creating pores with hydrophilic interiors. The effect is loss of cellular membrane potential and eventually cell death. CytK is a pore-forming protein more related to other hemolysins.

The timing of the toxin production was previously thought to be possibly responsible for the two different courses of disease, but in fact the emetic syndrome is caused by a toxin, cereulide, found only in emetic strains and is not part of the "standard toolbox" of *B. cereus*. Cereulide is a cyclic polypeptide containing three repeats of four amino acids: D-oxy-Leu—D-Ala—L-oxy-Val—L-Val (similar to valinomycin produced by *Streptomyces griseus*) produced by nonribosomal peptide synthesis. Cereulide is believed to bind to 5-hydroxytryptamine 3 (5-HT3) serotonin receptors, activating them and leading to increased afferent vagus nerve stimulation. It was shown independently by two research groups to be encoded on multiple plasmids: pCERE01 or pBCE4810. Plasmid pBCE4810 shares homology with the *Bacillus anthracis* virulence plasmid pXO1, which encodes the anthrax toxin. Periodontal isolates of *B. cereus* also possess distinct pXO1-like plasmids. Like most of cyclic peptides containing nonproteogenic amino acids, cereulid is resistant to heat, proteolysis, and acid conditions.

B. cereus is also known to cause difficult-to-eradicate chronic skin infections, though less aggressive than necrotizing fasciitis. *B. cereus* can also cause keratitis.

Diagnosis

In case of foodborne illness, the diagnosis of *B. cereus* can be confirmed by the isolation of more than 10^5 *B. cereus* organisms per gram from epidemiologically implicated food, but such testing is often not done because the illness is relatively harmless and usually self-limiting.

Identification Through Testing

Below is a list of differential techniques and results that can help to identify *Bacillus cereus* from other bacteria and *Bacillus* species.

- Anaerobic Growth: Positive
- Voges Proskauer test: Positive
- Acid produced from:
    - D-glucose: Positive
    - L-arabinose: Negative
    - D-xylose: Negative
    - D-mannitol: Negative
- Starch hydrolysis: Positive
- Nitrate reduction: Positive
- Degradation of tyrosine: Positive
- Growth at:
    - >50 °C: Negative
- Utilization of citrate: Positive

Prognosis

Most emetic patients recover within six to 24 hours, but in some cases, the toxin can be fatal. In 2014, 23 neonates receiving total parenteral nutrition contaminated with *B. cereus* developed septicaemia, with three of the infants later dying as a result of infection.

Bacteriophage

Bacteria of the *B. cereus* group are infected by bacteriophage belonging to the family Tectiviridae. This family includes tail-less phages that have a lipid membrane/ vesicle beneath the icosahedral protein shell and that are formed of approximately equal amounts of virus-encoded proteins and lipids derived from the host cell plasma membrane. Upon infection, the lipid membrane becomes a tail-like structure used in genome delivery. The genome is composed of ~15-kb linear double-stranded DNA (dsDNA) with long inverted terminal repeat sequences (100 bp). GIL01, Bam35, GIL16, AP50, and Wip1 are examples of temperate tectiviruses infecting the *B. cereus* group.

Yersinia Enterocolitica

Yersinia enterocolitica is a bacterium that causes an infectious disease called yersiniosis. Yersinia enterocolitica is a member of the Yersinia family of bacteria.

Yersiniosis is most often acquired by eating contaminated food, especially raw or undercooked pork products. The preparation of raw pork intestines (chitterlings) may be particularly risky. Infants can be infected if their caretakers handle raw chitterlings and then do not adequately clean their hands before handling the infant or the infant's toys, bottles, or pacifiers. Drinking contaminated unpasteurized milk or untreated water can also transmit the infection. Occasionally Y. enterocolitica infection occurs after contact with infected animals. On rare occasions, it can be transmitted as a result of the bacterium passing from the stools or soiled fingers of one person to the mouth of another person. This may happen when basic hygiene and handwashing habits are inadequate. Rarely, the organism is transmitted through contaminated blood during a transfusion.

Common symptoms of yersiniosis in children (who most often contract the disease) are fever, abdominal pain, and diarrhea, which is often bloody. Symptoms typically develop 4 to 7 days after exposure and may last 1 to 3 weeks or longer. In older children and adults, right-sided abdominal pain and fever may be the predominant symptoms and may be confused with appendicitis. In a small proportion of cases, complications such as skin rash, joint pains, or spread of bacteria to the bloodstream can occur.

Yersiniosis is generally diagnosed by detecting the organism in the stool. The organism can also be recovered from other sites, including the throat, lymph nodes, joint fluid, urine, bile, and blood.

Uncomplicated cases of diarrhea due to Y. enterocolitica usually resolve on their own without antibiotic treatment. However, in more severe or complicated infections, antibiotics such as aminoglycosides, doxycycline, trimethoprim-sulfamethoxazole, or fluoroquinolones may be useful.

Most infections with Y. enterocolitica are uncomplicated and resolve completely. Occasionally, some persons develop joint pain, most commonly in the knees, ankles or wrists. These joint pains usually develop about 1 month after the initial episode of diarrhea and generally resolve after 1 to 6 months. A skin rash, called "erythema nodosum," may also appear on the legs and trunk; this is more common in women. In most cases, erythema nodosum resolves spontaneously within a month.

To prevent the infection:

1. Avoid eating raw or undercooked pork.
2. Consume only pasteurized milk or milk products.
3. Wash hands with soap and water before eating and preparing food, after contact with animals, and after handling raw meat.
4. After handling raw chitterlings, clean hands and fingernails scrupulously with soap and water before touching infants or their toys, bottles, or pacifiers.

Someone other than the foodhandler should care for children while chitterlings are being prepared.

5. Prevent cross-contamination in the kitchen:
    - Use separate cutting boards for meat and other foods.
    - Carefully clean all cutting boards, counter-tops, and utensils with soap and hot water after preparing raw meat.
6. Dispose of animal feces in a sanitary manner.

Treatment

Yersiniosis is usually self-limiting and does not require treatment. For sepsis or severe focal infections, especially if associated with immunosuppression, the recommended regimen includes doxycycline in combination with an aminoglycoside. Other antibiotics active against *Y. enterocolitica* include trimethoprim-sulfamethoxasole, fluoroquinolones, ceftriaxone, and chloramphenicol. *Y. enterocolitica* is usually resistant to penicillin G, ampicillin, and cefalotin due to beta-lactamase production.

Staphylococcus Aureus

Staphylococcus aureus is a gram positive bacterium that usually appears under the microscope as spherical (coccus) organisms appearing in pairs, short chains, or bunched, grape-like clusters.

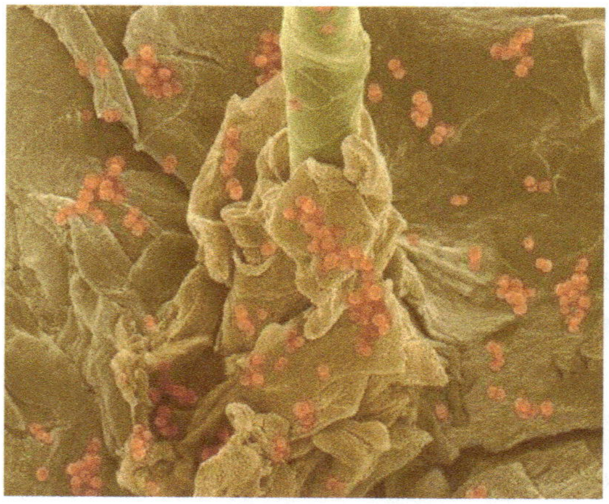

Dennis Kunkel Microscopy.

Diseases caused by Staphylococcus Aureus

Staphylococcus aureus has been implicated as a causative agent in acute food poisoning episodes, toxic shock syndrome, impetigo, scalded skin syndrome, cellulitis, folliculitis and furuncles.

It is also a common cause of systemic infections such as infective endocarditis, osteomyelitis, epiglottitis, and sinus infections amongst others. S. aureus is also responsible for many infective and systemic infections in the health care setting (nosocomial infections).

People Susceptible to Infections with Staphylococcus Aureus

All people are susceptible to infections with Staphylococcus aureus, the disease however, may manifest with varying intensities.

Epidemiology of Staphylococcus Aureus

Staphylococcus aureus can cause illness by preformed toxin production as well as by infecting both local tissues and the systemic circulation. Disease transmission can occur in gastrointestinal. Staphylococcus aureus causes acute episodes of food poisoning via preformed enterotoxins. Food items likely to be infected by staphylococcal food poisoning include meat and meat products; poultry and egg products; salads such as egg, tuna, chicken, potato, and macaroni; bakery products such as cream-filled pastries, cream pies, and chocolate éclairs; sandwich fillings; and milk and dairy products.

Salmonella spp.

Salmonella spp. are a group of bacteria which reside in the intestinal tract of human beings and warm blooded animals and are capable of causing disease. They are the second most common cause of bacterial foodborne illness in Ireland (*Campylobacter* spp. is the most frequent cause). They are facultative anaerobic Gramnegative rods.

Salmonella spp. are members of the *Enterobacteriaceae* group. The genus *Salmonella* contains 2 species:

- *Salmonella enterica*
- *Salmonella bongori*

Salmonella enterica is an important agent of foodborne illness. This species is sub-classified into 6 subspecies of which *S. enterica* subspecies enterica is the most important for human health.

The genus *Salmonella* can be subdivided into more than 2,400 serotypes. *Salmonella enterica* subsp. *Enterica* serotype Typhimurium (*S.* Typhimurium) and *Salmonella enterica* subsp. *enterica* serotype Enteritidis (*S.* Enteritidis) are the most frequently isolated serotypes in humans in Ireland. Serotypes are further subdivided by their resistance to bacteriophages (phage types or lystotypes), antibiotics or heavy metals; their biochemical characteristics (biovars or biotypes) or their sensitivity to or production of bacteriocins.

Growth and Survival Characteristics

Table: Factors affecting the growth of *Salmonella* spp.

Conditions	Minimum	Optimum	Maximum
Temperature (°C)	5.2	35 – 43	46.2
pH	3.8	7 – 7.5	9.5
Water activity (a_w)	0.94	0.99	>0.99

Salmonella spp. are not particularly heat resistant and most serotypes are killed by normal cooking conditions, i.e. cooking to a core temperature of 75°C instantaneously or an equivalent time temperature combination, e.g. 70°C for 2 minutes. However, a few highly heat resistant serotypes have been reported, e.g. *S.* Senftenberg 775W and *S.* Irumu. Heat resistance is influenced by water activity (aw), nature of the solutes and pH of the suspending medium. Greater heat resistance is observed for cells in sucrose compared with NaCl at the same aw values. Heat resistance also increases with increasing concentration of milk solids.

Heat resistance of *S.* Typhimurium in milk chocolate at various temperatures is presented in the following table:

	D-Value (Min) For S. Typhimurium In Milk Chocolate
90°C	75
80°C	222
70°C	816

Heat resistance is usually expressed by the D-value (i.e. the time in minutes at a given temperature to achieve a 90% reduction in the number of viable cells). ICMSF 1996 provides an extensive list of D-Values for *Salmonella* spp. under different conditions and food types.

Sources of *Salmonella* spp. and Routes of Transmission

Salmonella spp. reside in the intestinal tract of humans and warm blooded animals. They are shed in the faeces. Poor hygiene practices can result in the spread of this pathogen to the hands of humans while on the farm. Feet, hair and skin of animals can become contaminated as they walk, sit or lie in faecally contaminated ground or litter.

The spread of *Salmonella* spp. to carcasses may occur in the abattoir environment during slaughter. Furthermore, spread of *Salmonella* spp. to agricultural commodities may occur through faecal contamination of streams, rivers and coastal waters and the use of improperly treated sewage for agricultural purposes. In the food processing environment, spread of this pathogen can occur through cross contamination from raw food or infected food handlers.

The human infection caused by *Salmonella* spp. is referred to as salmonellosis. Although salmonellosis can arise from contact with infected animals, consumption of contaminated food is the most usual cause.

Salmonella spp. and Food

Many foods have been identified as vehicles for the transmission of this pathogen to humans, particularly foods of animal origin but also foods of non-animal origin which may be subject to faecal contamination. Those of particular importance include eggs, pig meat, poultry meat, milk, chocolate, fruit and vegetables. It is important to note that although *Salmonella* spp. do not form spores, they can survive for long periods in foods.

As pathogens

Salmonella species are facultative intracellular pathogens. A facultative organism uses oxygen to make ATP; when it is not available it "exercises its option"—the literal meaning of the term—and makes ATP via fermentation, or by substituting one or more of four less efficient electron acceptors as oxygen at the end of the electron transport chain: sulfate, nitrate, sulfur, or fumarate.

Most infections are due to ingestion of food contaminated by animal feces, or by human feces, such as by a food-service worker at a commercial eatery. *Salmonella* serotypes can be divided into two main groups—typhoidal and nontyphoidal. Nontyphoidal serotypes are more common, and usually cause self-limiting gastrointestinal disease. They can infect a range of animals, and are zoonotic, meaning they can be transferred between humans and other animals. Typhoidal serotypes include *Salmonella* Typhi and *Salmonella* Paratyphi A, which are adapted to humans and do not occur in other animals.

Nontyphoidal Salmonella

Infection with nontyphoidal serotypes of *Salmonella* generally results in food poisoning. Infection usually occurs when a person ingests foods that contain a high concentration of the bacteria. Infants and young children are much more susceptible to infection, easily achieved by ingesting a small number of bacteria. In infants, infection through inhalation of bacteria-laden dust is possible.

The organisms enter through the digestive tract and must be ingested in large numbers to cause disease in healthy adults. An infection can only begin after living salmonellae (not merely *Salmonella*-produced toxins) reach the gastrointestinal tract. Some of the microorganisms are killed in the stomach, while the surviving ones enter the small intestine and multiply in tissues. Gastric acidity is responsible for the destruction of the majority of ingested bacteria, but *Salmonella* has evolved a degree of tolerance to acidic environments that allows a subset of ingested bacteria to survive. Bacterial colonies may also become trapped in mucus produced in the esophagus. By the end of the incubation period, the nearby host cells are poisoned by endotoxins released from the dead salmonellae. The local response to the endotoxins is enteritis and gastrointestinal disorder.

About 2,000 serotypes of nontyphoidal *Salmonella* are known, which may be responsible for as many as 1.4 million illnesses in the United States each year. People who are at risk for severe illness include infants, elderly, organ-transplant recipients, and the immunocompromised.

Invasive Nontyphoidal Salmonella Disease

While in developed countries, nontyphoidal serotypes present mostly as gastrointestinal disease, in sub-Saharan Africa, these serotypes can create a major problem in bloodstream infections, and are the most commonly isolated bacteria from the blood of those presenting with fever. Bloodstream infections caused by nontyphoidal salmonellae in Africa were reported in 2012 to have a case fatality rate of 20–25%. Most cases of invasive nontyphoidal salmonella infection (iNTS) are caused by *S. typhimurium* or *S. enteritidis*. A new form of *Salmonella typhimurium* (ST313) emerged in the southeast of the African continent 75 years ago, followed by a second wave which came out of central Africa 18 years later. This second wave of iNTS possibly originated in the Congo Basin, and early in the event picked up a gene that made it resistant to the antibiotic chloramphenicol. This created the need to use expensive antimicrobial drugs in areas of Africa that were very poor, making treatment difficult. The increased prevalence of iNTS in sub-Saharan Africa compared to other regions is thought to be due to the large proportion of the African population with some degree of immune suppression or impairment due to the burden of HIV, malaria, and malnutrition, especially in children. The genetic makeup of iNTS is evolving into a more typhoid-like bacterium, able to efficiently spread around the human body. Symptoms are reported to be diverse, including fever, hepatosplenomegaly, and respiratory symptoms, often with an absence of gastrointestinal symptoms.

Typhoidal Salmonella

Typhoid fever is caused by *Salmonella* serotypes which are strictly adapted to humans or higher primates—these include *Salmonella* Typhi, Paratyphi A, Paratyphi B, and Paratyphi C. In the systemic form of the disease, salmonellae pass through the lymphatic system of the intestine into the blood of the patients (typhoid form) and are carried to various organs (liver, spleen, kidneys) to form secondary foci (septic form). Endotoxins first act on the vascular and nervous apparatus, resulting in increased permeability and decreased tone of the vessels, upset of thermal regulation, and vomiting and diarrhoea. In severe forms of the disease, enough liquid and electrolytes are lost to upset the water-salt metabolism, decrease the circulating blood volume and arterial pressure, and cause hypovolemic shock. Septic shock may also develop. Shock of mixed character (with signs of both hypovolemic and septic shock) is more common in severe salmonellosis. Oliguria and azotemia may develop in severe cases as a result of renal involvement due to hypoxia and toxemia.

Global Monitoring

In Germany, food poisoning infections must be reported. From 1990 to 2016, the number of officially recorded cases decreased from about 200,000 to about 13,000 cases. In the United States, about 1,200,000 cases of *Salmonella* infections are reported each year. A World Health Organization study estimated that 21,650,974 cases of typhoid fever occurred in 2000, 216,510 of which resulted in death, along with 5,412,744 cases of paratyphoid fever.

Molecular Mechanisms of Infection

Mechanisms of infection differ between typhoidal and nontyphoidal serotypes, owing to their different targets in the body and the different symptoms that they cause. Both groups must enter by crossing the barrier created by the intestinal cell wall, but once they have passed this barrier, they use different strategies to cause infection.

Nontyphoidal serotypes preferentially enter M cells on the intestinal wall by bacterial-mediated endocytosis, a process associated with intestinal inflammation and diarrhoea. They are also able to disrupt tight junctions between the cells of the intestinal wall, impairing the cells' ability to stop the flow of ions, water, and immune cells into and out of the intestine. The combination of the inflammation caused by bacterial-mediated endocytosis and the disruption of tight junctions is thought to contribute significantly to the induction of diarrhoea.

Salmonellae are also able to breach the intestinal barrier via phagocytosis and trafficking by CD18-positive immune cells, which may be a mechanism key to typhoidal *Salmonella* infection. This is thought to be a more stealthy way of passing the intestinal barrier, and may, therefore, contribute to the fact that lower numbers of typhoidal *Salmonella* are required for infection than nontyphoidal *Salmonella*. *Salmonella* cells are able to enter macrophages via macropinocytosis. Typhoidal serotypes can use this to achieve dissemination throughout the body via the mononuclear phagocyte system, a network of connective tissue that contains immune cells, and surrounds tissue associated with the immune system throughout the body.

Much of the success of *Salmonella* in causing infection is attributed to two type III secretion systems which function at different times during an infection. One is required for the invasion of nonphagocytic cells, colonization of the intestine, and induction of intestinal inflammatory responses and diarrhea. The other is important for survival in macrophages and establishment of systemic disease. These systems contain many genes which must work co-operatively to achieve infection.

The AvrA toxin injected by the SPI1 type III secretion system of *S.* Typhimurium works to inhibit the innate immune system by virtue of its serine/threonine acetyltransferase activity, and requires binding to eukaryotic target cell phytic acid (IP6). This leaves the host more susceptible to infection.

Salmonellosis is known to be able to cause back pain or spondylosis. It can manifest as five clinical patterns: gastrointestinal tract infection, enteric fever, bacteremia, local infection, and the chronic reservoir state. The initial symptoms are nonspecific fever, weakness, and myalgia among others. In the bacteremia state, it can spread to any parts of the body and this induces localized infection or it forms abscesses. The forms of localized *Salmonella* infections are arthritis, urinary tract infection, infection of the central nervous system, bone infection, soft tissue infection, etc. Infection may remain as the latent form for a long time, and when the function of reticular endothelial cells is deteriorated, it may become activated and consequently, it may secondarily induce spreading infection in the bone several months or several years after acute salmonellosis.

Resistance to Oxidative Burst

A hallmark of *Salmonella* pathogenesis is the ability of the bacterium to survive and proliferate within phagocytes. Phagocytes produce DNA damaging agents such as nitric oxide and oxygen radicals as a defense against pathogens. Thus, *Salmonella* species must face attack by molecules that challenge genome integrity. Buchmeier *et al.* showed that mutants of *S. enterica* lacking RecA or RecBC protein function are highly sensitive to oxidative compounds synthesized by macrophages, and furthermore these findings indicate that successful systemic infection by *S. enterica* requires RecA and RecBC mediated recombinational repair of DNA damage.

Host Adaptation

S. enterica, through some of its serotypes such as Typhimurium and Enteriditis, shows signs of the ability to infect several different mammalian host species, while other serotypes such as Typhi seem to be restricted to only a few hosts. Some of the ways that *Salmonella* serotypes have adapted to their hosts include loss of genetic material and mutation. In more complex mammalian species, immune systems, which include pathogen specific immune responses, target serovars of *Salmonella* through binding of antibodies to structures like flagella. Through the loss of the genetic material that codes for a flagellum to form, *Salmonella* can evade a host's immune system. *mgtC* leader RNA from bacteria virulence gene (mgtCBR operon) decreases flagellin production during infection by directly base pairing with mRNAs of the *fljB* gene encoding flagellin and promotes degradation. In the study by Kisela *et al.*, more pathogenic serovars of *S. enterica* were found to have certain adhesins in common that have developed out of convergent evolution. This means that, as these strains of *Salmonella* have been exposed to similar conditions such as immune systems, similar structures evolved separately to negate these similar, more advanced defenses in hosts. There are still many questions about the way that *Salmonella* has evolved into so many different types but it has been suggested that *Salmonella* evolved through several phases. As Baumler *et al.* have suggested, Salmonella most likely evolved through horizontal gene transfer, formation of new serovars due to additional pathogenicity islands and through an approximation of its ancestry. So, *Salmonella* could have evolved into its many different serotypes through gaining genetic information from different pathogenic bacteria. The presence of several pathogenicity islands in the genome of different serotypes has lent credence to this theory.

Genetics

In addition to its importance as a pathogen, *S. enterica* serovar Typhimurium has been instrumental in the development of genetic tools that led to an understanding of fundamental bacterial physiology. These developments were enabled by the discovery of the first generalized transducing phage, P22, in Typhimurium that allowed quick and easy genetic exchange that allowed fine structure genetic analysis. The large number of mutants led to a revision of genetic nomenclature for bacteria. Many of the uses of transposons as genetic tools, including transposon delivery, mutagenesis, construction of chromosome rearrangements, were also developed in Typhimurium. These genetic tools also led to a simple test for carcinogens, the Ames Test.

Vibrio spp.

Vibrio, (genus Vibrio), any of a group of comma-shaped bacteria in the family Vibrionaceae. Vibrios are aquatic microorganisms, some species of which cause serious diseases in humans and other animals.

Vibrios are microbiologically characterized as gram-negative, highly motile, facultative anaerobes (not requiring oxygen), with one to three whiplike flagella at one end. Their cells are curved rods 0.5 μm (micrometre; 1 μm = 10^{-6} metre) across and 1.5 to 3.0 μm long, single or strung together in S-shapes or spirals.

Three species of vibrio are of significance to humans: V. cholerae is the cause of cholera, and V. parahaemolyticus and V. vulnificus both act as agents of acute enteritis, or bacterial diarrhea. V. anguillarum is found in diseased eels and other fishes.

Ways People Get Vibrosis

Most people become infected by eating raw or undercooked shellfish, particularly oysters. Certain Vibrio species can also cause a skin infection when an open wound is exposed to brackish or salt water. Brackish water is a mixture of fresh and sea water. It is often found where rivers meet the sea.

People Prone to Vibriosis

People with compromised immune systems, especially those with chronic liver disease, are more likely to get vibriosis. Eating raw seafood, particularly oysters, and exposing open wounds to brackish or salt water can increase a person's chance for getting vibriosis. About 80% of infections occur between May and October when water temperatures are warmer.

How Common is Vibriosis?

It is estimated that vibriosis causes 80,000 illnesses each year in the United States. About 52,000 of these illnesses are estimated to be the result of eating contaminated food.

The most commonly reported species, Vibrio parahaemolyticus, is estimated to cause 45,000 illnesses each year in the United States.

Is Vibriosis a Serious Disease?

Most people with a mild case of vibriosis recover after about 3 days with no lasting effects. However, people with a Vibrio vulnificus infection can get seriously ill and need intensive care or limb amputation. About 1 in 4 people with this type of infection die, sometimes within a day or two of becoming ill.

Prevention of Vibriosis

To reduce your chance of getting vibriosis, don't eat raw or undercooked shellfish, such as oysters. If you have a wound (including cuts and scrapes), avoid contact with brackish or salt water or cover the wound with a waterproof bandage if there's a possibility it could come into contact with brackish or salt water, raw seafood, or raw seafood juices.

Clostridium spp.

It is a microbe that lives right here on earth. This microbe is a bacterium that is soil born and can be found in the intestines of animals.

Clostridium sporogenes, or C. sporogenes for short, is a Gram-positive bacteria. What is that you might ask? A Gram-positive bacterium is one that stains violet in color during a Gram stain test. This means that it has a very thick outer cell wall made up of peptidoglycan, which is a sugar structure that creates a protective layer around the bacteria to keep is safe from attack.

Clostridia are gram-positive spore-forming bacilli that cause up to 10% of anaerobic infections and occur at diverse clinical sites. Of the more than 200 recognized species of Clostridium, <20 are associated with invasive and toxigenic human diseases. Clostridial species vary in oxygen tolerance, nutritional needs, motility, and optimal temperature for growth. They can cause distinct clostridial histotoxic syndromes due to specific clostridial toxins (e.g., gas gangrene, food poisoning) and other, nonsyndromic clostridial infections (e.g., bloodstream infection, BSI). Clostridia also are recovered from various sites as part of polymicrobial infection, including intra-abdominal, biliary tract, female genital tract, pleuropulmonary, central nervous system, and skin and soft-tissue infections.

Clostridium contains around 100 species that include common free-living bacteria, as well as important pathogens. The main species responsible for disease in humans are:

- *Clostridium botulinum* can produce botulinum toxin in food or wounds and can cause botulism. This same toxin is known as Botox and is used in cosmetic surgery to paralyze facial muscles to reduce the signs of aging; it also has numerous other therapeutic uses.
- *Clostridium difficile* can flourish when other gut flora bacteria are killed during antibiotic therapy, leading to superinfection and potentially fatal pseudomembranous colitis (a severe necrotizing disease of the large intestine).
- *Clostridium perfringens* causes a wide range of symptoms, from food poisoning to cellulitis, fasciitis, and gas gangrene.
- *Clostridium tetani* causes tetanus.
- *Clostridium sordellii* can cause a fatal infection in exceptionally rare cases after medical abortions.

Bacillus and *Clostridium* are often described as gram-variable, because they show an increasing number of gram-negative cells as the culture ages.

Clostridium and *Bacillus* are both in the division Firmicutes, but they are in different classes, orders, and families. Microbiologists distinguish *Clostridium* from *Bacillus* by the following features:

- *Clostridium* grows in anaerobic conditions; *Bacillus* grows in aerobic conditions.
- *Clostridium* forms bottle-shaped endospores; *Bacillus* forms oblong endospores.
- *Clostridium* does not form the enzyme catalase; *Bacillus* secretes catalase to destroy toxic byproducts of oxygen metabolism.

Clostridium and *Desulfotomaculum* are both in the class Clostridia and order Clostridiales, and they both produce bottle-shaped endospores, but they are in different families. *Clostridium* can be distinguished from *Desulfotomaculum* on the basis of the nutrients each genus uses (the latter requires sulfur).

Glycolysis and fermentation of pyruvic acid by Clostridia yield the end products butyric acid, butanol, acetone, isopropanol, and carbon dioxide.

The Schaeffer-Fulton stain (0.5% malachite green in water) can be used to distinguish endospores of *Bacillus* and *Clostridium* from other microorganisms. There is a commercially available polymerase chain reaction (PCR) test kit (Bactotype) for the detection of *C. perfringens* and other pathogenic bacteria.

Treatment

In general, the treatment of clostridial infection is high-dose penicillin G, to which the organism has remained susceptible. *Clostridium welchii* and *Clostridium tetani* respond to sulfonamides. Clostridia are also susceptible to tetracyclines, carbapenems (imipenem), metronidazole, vancomycin, and chloramphenicol.

The vegetative cells of clostridia are heat-labile and are killed by short heating at temperatures above 72–75 °C. The thermal destruction of *Clostridium* spores requires higher temperatures (above 121.1 °C, for example in an autoclave) and longer cooking times (20 min, with a few exceptional cases of > 50 min recorded in the literature). *Clostridia* and *Bacilli* are quite radiation-resistant, requiring doses of about 30 kGy, which is a serious obstacle to the development of shelf-stable irradiated foods for general use in the retail market. The addition of lysozyme, nitrate, nitrite and propionic acid salts inhibits clostridia in various foods.

Fructooligosaccharides (fructans) such as inulin, occurring in relatively large amounts in a number of foods such as chicory, garlic, onion, leek, artichoke, and asparagus, have a prebiotic or bifidogenic effect, selectively promoting the growth and metabolism of beneficial bacteria in the colon, such as bifidobacteria and lactobacilli, while inhibiting harmful ones, such as clostridia, fusobacteria, and bacteroides.

Clostridium Botulinum

Botulism is a rare but potentially life-threatening bacterial illness. *Clostridium Botulinum* bacteria grows on food and produces toxins that, when ingested, cause paralysis. Botulism poisoning is extremely rare, but so dangerous that each case is considered a public health emergency. Studies have shown that there is a 35 to 65 percent chance of death for patients who are not treated immediately and effectively with botulism antitoxin. Infant botulism is the most common form of botulism.

Most of the botulism cases reported each year come from foods that are not canned properly at home. Botulism from commercially canned food is rare, but commercial canned chili products were identified as the source of a botulism outbreak in 2007.

Symptoms of Botulism

Botulism neurotoxins prevent neurotransmitters from functioning properly. This means that they inhibit motor control. As botulism progresses, the patient experiences paralysis from top to bottom, starting with the eyes and face and moving to the throat, chest, and extremities. When paralysis reaches the chest, death from inability to breathe results unless the patient is ventilated. Symptoms of botulism generally appear 12 to 72 hours after eating contaminated food. With treatment, illness lasts from 1 to 10 days. Full recovery from botulism poisoning can take weeks to months. Some people never fully recover.

In general, symptoms of botulism poisoning include the following:

- Nausea
- Vomiting
- Fatigue
- Dizziness
- Double vision
- Dry skin, mouth and throat
- Drooping eyelids
- Difficulty swallowing
- Slurred speech
- Muscle Weakness
- Body Aches
- Paralysis
- Lack of fever

Infant botulism takes on a different form. Symptoms in an infant include lethargy, poor appetite, constipation, drooling, drooping eyelids, a weak cry, and paralysis.

Long-Term Effects of Botulism

The majority of botulism patients never fully recover their pre-illness health. After three months to a year of recovery, persisting side-effects are most likely permanent. These long-term effects most often include fatigue, weakness, dizziness, dry mouth, and difficulty performing strenuous tasks. Patients also report a generally less happy and peaceful psychological state than before their illness.

Botulism Diagnosis

If a patient displays symptoms of botulism, a doctor will most likely take a blood, stool, or gastric secretion sample. The most common test for botulism is injecting the patient's blood into a mouse to see whether the mouse displays signs of botulism, since other testing methods take up to a week.

Sometimes botulism can be difficult to diagnose, since symptoms can be mild, or confused with those of Guillan-Barre Syndrome.

Treatment of Botulism

If found early, botulism can be treated with an antitoxin that blocks circulation of the toxin in the bloodstream. This prevents the patient's case from worsening, but recovery still takes several weeks.

Clostridium Perfringens

Clostridium perfringens (formerly known as *C. welchii*, or *Bacillus welchii*) is a Gram-positive, rod-shaped, anaerobic, spore-forming pathogenic bacterium of the genus *Clostridium*. *C. perfringens* is everpresent in nature and can be found as a normal component of decaying vegetation, marine sediment, the intestinal tract of humans and other vertebrates, insects, and soil. It has the shortest reported generation time of any organism at 6.3 minutes in thioglycolate medium.

C. perfringens is one of the most common causes of food poisoning in the United States, alongside norovirus, *Salmonella*, *Campylobacter*, and *Staphylococcus aureus*. However, it can sometimes be ingested and cause no harm.

Infections due to *C. perfringens* show evidence of tissue necrosis, bacteremia, emphysematous cholecystitis, and gas gangrene, which is also known as clostridial myonecrosis. The toxin involved in gas gangrene is known as α-toxin, which inserts into the plasma membrane of cells, producing gaps in the membrane that disrupt normal cellular function. *C. perfringens* can participate in polymicrobial anaerobic infections. *Clostridium perfringens* is commonly encountered in infections as a component of the normal flora. In this case, its role in disease is minor.

The action of *C. perfringens* on dead bodies is known to mortuary workers as tissue gas and can be halted only by embalming.

Genome

C. perfringens has a stable G+C content of approximately 27–28% and average genome size of 3.5 Mb. Genomes of 56 *C. perfringens* strains have since been made available on NCBI genomes database for scientific research community. Genomic research has revealed surprisingly high diversity in *C. perfringens* pangenome, with only 12.6% core genes, identified as the most divergent Gram-positive bacteria reported. Nevertheless, 16S rRNA regions in between *C. perfringens* strains are found to be highly conserved (sequence identity >99.1%).

Motility

Although they lack flagella, *C. perfringens* bacteria are able to glide across surfaces because their bodies are lined with filaments from end-to-end. The hypermotile variants such as SM101, are often found arising on the edges of colonies on agar plates. Video microscopy of their gliding movement suggest that they form long, thin filaments that allow them to move rapidly like bacteria with flagella. Genome sequencing was used to identify the cause(s) of the hypermotile phenotype and their direct derivatives. In comparing them, strains SM124 and SM127, hypermotile derivatives of strains SM101 and SM102, respectively, contained 10 and 6 nucleotide polymorphisms (SNPs) relative to their parent strains. Mutations in cell division genes is the common feature of the hypermotile strains.

Food Poisoning

In the United Kingdom and United States, *C. perfringens* bacteria are the third most common cause of foodborne illness, with poorly prepared meat and poultry, or food properly prepared but left to stand too long, the main culprits in harboring the bacterium. The clostridium perfringens enterotoxin (CPE) mediating the disease is heat-labile (inactivated at 74 °C (165 °F)). It can be detected in contaminated food (if not heated properly), and feces. Incubation time is between six and 24 (commonly 10–12) hours after ingestion of contaminated food.

Since *C. perfringens* forms spores that can withstand cooking temperatures, if cooked food is let stand for long enough, germination can ensue and infective bacterial colonies develop. Symptoms typically include abdominal cramping, diarrhea, vomiting, and fever. The whole course usually resolves within 24 hours, but can last up to two weeks in older or infirm hosts.

Perfringens poisoning can also lead to another disease known as enteritis necroticans (also known as pigbel); this is caused by *C. perfringens* type c. However this infection is often fatal. Large numbers of *C. perfringens* grow in the intestines, and secrete exotoxin. This exotoxin causes necrosis of the intestines, varying levels of hemorrhaging and perforation of the intestine. Inflammation usually occurs in sections of the jejunum, midsection of the small intestine. This disease eventually leads to septic shock and death. This particular disease is rare in the United States; typically it occurs in populations with a higher risk. Risk factors for enteritis necroticans include protein-deficient diet, unhygienic food preparation, sporadic feasts of meat (after long periods of a protein-deficient diet), diets containing large amounts of trypsin inhibitors (sweet potatoes), areas prone to infection of the parasite Ascaris (produces a trypsin inhibitor). This disease is contracted in populations living in New Guinea, parts of Africa, Central America, South America and Asia.

Many cases of *C. perfringens* food poisoning likely remain subclinical, as antibodies to the toxin are common among the population. This has led to the conclusion that most of the population has experienced food poisoning due to *C. perfringens*.

Despite its potential dangers, *C. perfringens* is used as the leavening agent in salt rising bread. The baking process is thought to reduce the bacterial contamination, precluding negative effects.

Infection

Clostridium perfringens is the most common bacterial agent for gas gangrene. Some symptoms include blisters, tachycardia, swelling, and jaundice.

Research published in 2014 suggested that a strain of *C. perfringens* might be implicated in multiple sclerosis (MS). Tests in mice found that a toxin made by a rare strain of *C. perfringens* caused MS-like damage in the brain, and earlier work had identified this strain of *C. perfringens* in a human with MS. MS patients were found to be ten times more immune-reactive to the epsilon toxin than healthy people.

Diagnosis

C. perfringens can be diagnosed by Nagler's reaction where the suspect organism is cultured on an egg yolk media plate. One side of the plate contains anti-alpha-toxin, while the other side does not. A streak of suspect organism is placed through both sides. An area of turbidity will form around the side that does not have the anti-alpha-toxin, indicating uninhibited lecithinase activity. In addition, laboratories can diagnose the bacteria by determining the number of bacteria in the feces. Within the 48 hours from when the disease began, if the individual has more than 106 spores of the bacteria per gram of stool, then the illness is diagnosed as *C. perfringens* food poisoning. Other tests/reactions: Catalase: Negative, Spot indole: Positive, Lecithinase: Positive, Lipase: Negative, Litmus Milk: Stormy Fermentation, Reverse CAMP plate: Positive, Gas Liquid Chromatography products: (Acetic, Butyric and Lactic Acids).

Typically, the symptoms of *C. perfingens* poisoning are used to diagnose it. However, diagnosis can be made using a stool culture test, in which the feces is tested for toxins produced by the bacteria.

Prevention

The growth of *C. perfringens* spores can be prevented by most importantly cooking food, especially beef and poultry, thoroughly to the recommended temperatures. Leftover food should be refrigerated to a temperature below 40 °F (4 °C) within two hours of preparation. Large pots of food like soup or stew with meats should be divided in small quantities and covered for refrigeration. Leftovers should be reheated to at least 165 °F (74 °C) before serving. A rule of thumb is that if the food tastes, smells, or looks different from what it is supposed to then the food should be avoided. Even if it looks okay, a food that has been out for a long time can also be dangerous to eat.

Food Poisoning Incidents

On May 7, 2010, "42 residents and 12 staff members at a Louisiana state psychiatric hospital were affected experienced vomiting, abdominal cramps, and diarrhea". Three patients died within 24 hours. The outbreak was linked to chicken which was cooked a day prior to being served and was not cooled down according to hospital guidelines. The outbreak affected 31% of the residents of the hospital and 69% of the staff who ate the chicken. It is unknown how many of the affected residents ate the chicken.

In May 2011, a man died after allegedly eating food contaminated with the bacteria on a transatlantic American Airlines flight. The man's wife and daughter are suing American and LSG Sky Chefs, the German company that prepared the inflight food.

In December 2012, a 46-year-old woman died two days after eating a Christmas Day meal at a pub in Hornchurch, Essex, England. She was among about 30 people to fall ill after eating the meal. Samples taken from the victims contained *C. perfringens*. The hotel manager and the cook were jailed for offences arising from the incident.

In December 2014, 87-year-old Bessie Scott died three days after eating a church pot-luck supper in Nackawic, New Brunswick, Canada. Over 30 other people reported signs of gastrointestinal illness, diarrhea and abdominal pain. The province's acting chief medical officer says *Clostridium perfringens* is the bacteria that most likely caused the woman's death.

In October 2016, 66 year-old Alex Zdravich died four days after eating an enchilada, burrito, and taco at Agave Azul in West Lafayette, Indiana. Three others that dined the same day reported signs of foodborne illness, which were consistent with the symptoms and rapid onset of *Clostridium perfringens*. They later tested positive for the presence of the bacteria.

In November 2016, food contaminated with *Clostridium perfringens* caused three individuals to die, and another 22 to be sickened, after a Thanksgiving luncheon hosted by a church in Antioch, California.

In January 2017, a mother and her son sued a restaurant in Rochester, New York as they and 260 other people were sickened after eating foods contaminated with

Clostridium perfringens. "Officials from the Monroe County Department of Public Health closed down the Golden Ponds after more than a fourth of its Thanksgiving Day guests became ill. An inspection revealed a walk-in refrigerator with food spills and mold, a damaged gasket preventing the door from closing, and mildew growing inside."

Epidemiology

C. perfringens is the leading cause of food poisoning in the United States and Canada. Contaminated meats in stews, soups and gravies are usually responsible for outbreaks and causes about 250,000 cases of foodborne illnesses in the United States every year. Deaths due to the disease are rare and mostly occurs in elderly and people who are predisposed to the disease. From 1998–2010, 289 confirmed outbreaks of *C. perfringens* illness were reported with 15,208 illnesses, 82 hospitalizations, and 8 deaths. Outbreaks can cause regularly and can cause substantial morbidity but it can be prevented by avoiding the contamination of meats that were coming from the slaughterhouses and farms.

Shigella spp.

Shigella is the bacterium that causes the disease shigellosis, also known as bacillary dysentery. *Shigella* is one of the most easily transmitted bacterial diarrheas, since it can occur after fewer than 100 bacteria are ingested. While reported cases of *Shigella* range between 14,000 and 20,000 annually, with the majority of these cases occurring between July and October. *Shigella Sonnei* is the most common type of *Shigella*. It accounts for over two-thirds of cases of shigellosis in the United States.

Shigella bacteria are generally transmitted through a fecal-oral route. Foods that come into contact with human or animal waste can transmit *Shigella*. Thus, handling toddlers' diapers, eating vegetables from a field contaminated with sewage, or drinking pool water are all activities that can lead to shigellosis.

Symptoms of Shigella Food Poisoning

Symptoms of *Shigella* poisoning most commonly develop one to three days after exposure to *Shigella* bacteria, and usually go away within five to seven days. It is also possible to get *Shigella* but experience no symptoms, and still be contagious to others, a condition known as being asymptomatic.

Common *Shigella* Food Poisoning Symptoms:

- Diarrhea: Diarrhea ranges from mild to severe. It is bloody in 25 to 50 percent of cases and usually contains mucus.
- Fever.

- Stomach cramps.
- Rectal spasms.

Complications from Shigella

Complications from shigellosis can include severe dehydration, seizures in small children, rectal bleeding, and invasion of the blood stream by the bacteria. Young children and the elderly are at the highest risk of death. The following is a list of specific complications caused by *Shigella*.

Proctitis and Rectal Prolapse: The bacteria that causes shingellosis can also cause inflammation of the lining of the rectum or rectal prolapse.

Reactive Arthritis: Approximately 3 percent of patients with *Shigella* infection, most often those with *Shigella flexneri*, develop Reactive Arthritis. It occurs when the immune system attempts to combat *Shigella* but instead attacks the body. Symptoms of Reactive Arthritis include inflammation of the joints, eyes, or reproductive or urinary organs. On average, symptoms appear 18 days after infection.

Toxic Megacolon: In this rare complication, the colon is paralyzed and unable to pass bowel movements or gas. Symptoms of Toxic Megacolon include abdominal pain and swelling, fever, weakness, and disorientation. If this complication goes untreated and the colon ruptures, the patient's condition can be life-threatening.

Hemolytic Uremic Syndrome, or HUS: *Shigella* rarely results in HUS, which is more commonly a complication of Shiga toxin-producing E. coli infections. HUS can lead to kidney failure.

Diagnosis of Shigella

A *Shigella* infection is diagnosed through laboratory testing of a stool sample.

Shigella Food Poisoning Treatment

A *Shigella* infection usually goes away on its own in five to seven days, although bowel movements may continue to be abnormal for up to a month following infection. Antibiotics, however, can shorten the course of the illness. A doctor can prescribe antibiotics after testing a stool sample for the presence of *Shigella* bacteria.

Some strains of shigellosis are resistant to antibiotics, meaning that antibiotics might not always be an effective treatment. Antidiarrheal medication should be avoided, as it can actually make the illness worse.

Listeria Monocytogenes

Listeria monocytogenes is the species of pathogenic bacteria that causes the infection listeriosis. It is a facultative anaerobic bacterium, capable of surviving in the presence or absence of oxygen. It can grow and reproduce inside the host's cells and is one of the most virulent foodborne pathogens, with 20 to 30% of food borne listeriosis infections in high-risk individuals may be fatal. Responsible for an estimated 1,600 illnesses and 260 deaths in the United States (U.S.) annually, listeriosis ranks third in total number of deaths among food borne bacterial pathogens, with fatality rates exceeding even *Salmonella* and *Clostridium botulinum*. In the European Union listeriosis follows an upward trend that began in 2008, causing 2,161 confirmed cases and 210 reported deaths in 2014, 16% more than in 2013. Listeriosis mortality rates in the US are also higher in the EU than for other food-borne pathogens.

Listeria monocytogenes is a Gram-positive bacterium, in the division Firmicutes, named after Joseph Lister. Its ability to grow at temperatures as low as 0 °C permits multiplication at typical refrigeration temperatures, greatly increasing its ability to evade control in human foodstuffs. Motile via flagella at 30 °C and below, but usually not at 37 °C, *L. monocytogenes* can instead move within eukaryotic cells by explosive polymerization of actin filaments (known as comet tails or actin rockets).

Studies suggest up to 10% of human gastrointestinal tracts may be colonized by *Listeria monocytogenes*.

Nevertheless, clinical diseases due to *Listeria monocytogenes* are more frequently recognized by veterinarians, especially as meningoencephalitis in ruminants.

Due to its frequent pathogenicity, causing meningitis in newborns (acquired transvaginally), pregnant mothers are often advised not to eat soft cheeses such as Brie, Camembert, feta, and queso blanco fresco, which may be contaminated with and permit growth of *Listeria monocytogenes*. It is the third-most-common cause of meningitis in newborns. *Listeria monocytogenes* can infect the brain, spinal cord membranes and/or the bloodstream of the host through the ingestion of contaminated food such as unpasteurized dairy or raw foods.

Classification

Listeria monocytogenes is a Gram-positive, non spore-forming, motile, facultatively anaerobic, rod-shaped bacterium. It is catalase-positive and oxidase-negative, and expresses a beta hemolysin, which causes destruction of red blood cells. This bacterium exhibits characteristic tumbling motility when viewed with light microscopy. Although *L. monocytogenes* is actively motile by means of peritrichous flagella at room temperature (20–25 °C), the organism does not synthesize flagella at body temperatures (37 °C).

The genus *Listeria* belongs to the class Bacilli and the order Bacillales, which also includes *Bacillus* and *Staphylococcus*. *Listeria* currently contains 10 species: *L. fleischmannii, L. grayi, L. innocua, L. ivanovii, L. marthii, L. monocytogenes,*

L. rocourtiae, L. seeligeri, L. weihenstephanensis and *L. welshimeri. L. denitrificans*, previously thought to be part of the genus *Listeria*, was reclassified into the new genus *Jonesia*. Both *L. ivanovii* and *L. monocytogenes* are pathogenic in mice, but only *L. monocytogenes* is consistently associated with human illness. The 13 serotypes of *L. monocytogenes* can cause disease, but more than 90% of human isolates belong to only three serotypes: 1/2a, 1/2b, and 4b. *L. monocytogenes* serotype 4b strains are responsible for 33 to 5% of sporadic human cases worldwide and for all major foodborne outbreaks in Europe and North America since the 1980s.

Pathogenesis

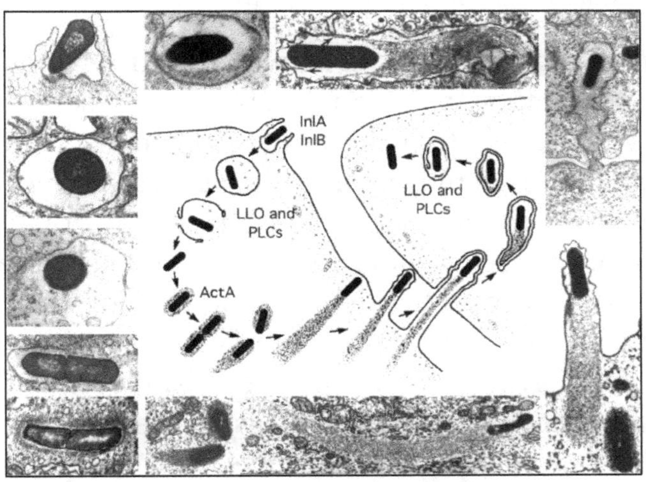

Stages in the intracellular lifecycle of *L. monocytogenes*. (Center) Cartoon depicting entry, escape from a vacuole, actin nucleation, actin-based motility, and cell-to-cell spread. (Outside) Representative electron micrographs from which the image was derived.

Invasive infection by *L. monocytogenes* causes the disease listeriosis. When the infection is not invasive, any illness as a consequence of infection is termed febrile gastroenteritis. The manifestations of listeriosis include septicemia, meningitis (or meningoencephalitis), encephalitis, corneal ulcer, pneumonia, and intrauterine or cervical infections in pregnant women, which may result in spontaneous abortion (second to third trimester) or stillbirth. Surviving neonates of fetomaternal listeriosis may suffer granulomatosis infantiseptica — pyogenic granulomas distributed over the whole body — and may suffer from physical retardation. Influenza-like symptoms, including persistent fever, usually precede the onset of the aforementioned disorders. Gastrointestinal symptoms, such as nausea, vomiting, and diarrhea, may precede more serious forms of listeriosis or may be the only symptoms expressed. Gastrointestinal symptoms were epidemiologically associated with use of antacids or cimetidine. The onset time to serious forms of listeriosis is unknown, but may range from a few days to three weeks. The onset time to gastrointestinal symptoms is unknown but probably exceeds 12 hours. An early study suggested that *L. monocytogenes* is unique among Gram-positive bacteria in that it might possess lipopolysaccharide, which serves as an endotoxin. Later, it was found to not be a true endotoxin. *Listeria* cell walls consistently contain lipoteichoic acids, in which a glycolipid moiety, such as a galactosyl-glucosyl-diglyceride, is covalently linked to the terminal phosphomonoester of the teichoic acid. This lipid region anchors the polymer chain to the cytoplasmic membrane. These lipoteichoic acids resemble the lipopolysaccharides of Gram-negative bacteria in both structure and function, being the only amphipathic polymers at the cell surface.

L. monocytogenes has D-galactose residues on its surface that can attach to D-galactose receptors on the host cell walls. These host cells are generally M cells and Peyer's patches of the intestinal mucosa. Once attached to this cells, *L. monocytogenes* can translocate past the intestinal membrane and into the body.

The infective dose of *L. monocytogenes* varies with the strain and with the susceptibility of the victim. From cases contracted through raw or supposedly pasteurized milk, one may safely assume that, in susceptible persons, fewer than 1,000 total organisms may cause disease. *L. monocytogenes* may invade the gastrointestinal epithelium. Once the bacterium enters the host's monocytes, macrophages, or polymorphonuclear leukocytes, it becomes bloodborne (septicemic) and can grow. Its presence intracellularly in phagocytic cells also permits access to the brain and probably transplacental migration to the fetus in pregnant women. The pathogenesis of *L. monocytogenes* centers on its ability to survive and multiply in phagocytic host cells. It seems that *Listeria* originally evolved to invade membranes of the intestines, as an intracellular infection, and developed a chemical mechanism to do so. This involves a bacterial protein "internalin"(InlA/InlB) which attaches to a protein on the intestinal cell membrane "cadherin" and allows the bacteria to invade the cells through a zipper mechanism. These adhesion molecules are also to be found in two other unusually tough barriers in humans — the blood-brain barrier and the fetal–placental barrier, and this may explain the apparent affinity that *Listeria* has for causing meningitis and affecting babies *in utero*. Once inside the cell, Listeria rapidly acidifies the lumen of the vacuole formed around it during cell entry to activate listeriolysin O, a cholesterol-dependent cytolysin capable of disrupting the vacuolar membrane. This frees the pathogen and gives it access to the cytosol of the cell, where it continues its pathogenesis. Motility in the intracellular space is provided by actin assembly-inducing protein (ActA) which allows the bacteria to use the host cell's actin polymerization machinery to polymerize the cytoskeleton to give a "boost" to the bacterial cell so it can move in the cell. The same ActA mechanism also allows the bacteria to travel from cell to cell.

Regulation of Pathogenesis

L. monocytogenes can act as a saprophyte or a pathogen, depending on its environment. When this bacterium is present within a host organism, quorum sensing causes the up-regulation of several virulence genes. Depending on the location of the bacterium within the host organism, different activators up-regulate the virulence genes. SigB, an alternative sigma factor, up-regulates *Vir* genes in the intestines, whereas PrfA up-regulates gene expression when the bacterium is present in blood. Little is known about how this bacterium switches between acting as a saprophyte and a pathogen; however, several noncoding RNAs are thought to be required to induce this change.

Pathogenicity of Lineages

L. monocytogenes has three distinct lineages, with differing evolutionary histories and pathogenic potentials. Lineage I strains contain the majority of human clinical isolates and all human epidemic clones, but are underrepresented in animal clinical isolates. Lineage II strains are overrepresented in animal cases and underrepresented in human clinical cases, and are more prevalent in environmental and food samples. Lineage III isolates are very rare, but significantly more common in animal than human isolates.

Detection

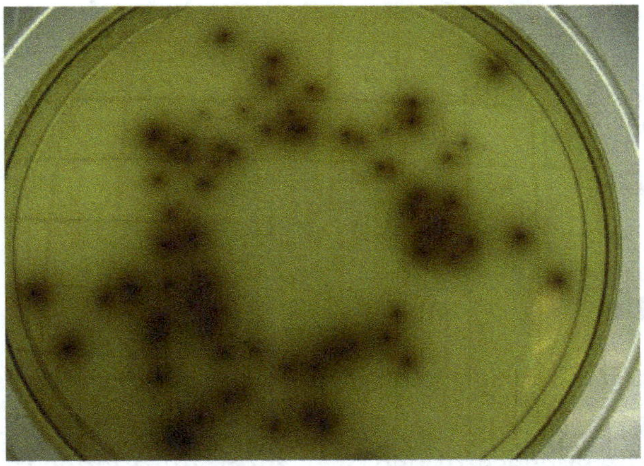

Colonies of typical *L. monocytogenes* as they appear when grown on *Listeria*-selective agar.

The Anton test is used in the identification of *L. monocytogenes*; instillation of a culture into the conjunctival sac of a rabbit or guinea pig causes severe keratoconjunctivitis within 24 hours.

Listeria species grow on media such as Mueller-Hinton agar. Identification is enhanced if the primary cultures are done on agar containing sheep blood, because the characteristic small zone of hemolysis can be observed around and under colonies. Isolation can be enhanced if the tissue is kept at 4 °C for some days before inoculation into bacteriologic media. The organism is a facultative anaerobe and is catalase-positive and motile. *Listeria* produces acid, but not gas, in a variety of carbohydrates. The motility at room temperature and hemolysin production are primary findings that help differentiate listeria from coryneform bacteria.

The methods for analysis of food are complex and time-consuming. The present U.S. FDA method, revised in September 1990, requires 24 and 48 hours of enrichment, followed by a variety of other tests. Total time to identification takes five to seven days, but the announcement of specific nonradiolabelled DNA probes should soon allow a simpler and faster confirmation of suspect isolates.

Recombinant DNA technology may even permit two- to three-day positive analysis in the future. Currently, the FDA is collaborating in adapting its methodology to quantitate very low numbers of the organisms in foods.

Treatment

When listeric meningitis occurs, the overall mortality may reach 70%, from septicemia 50%, and from perinatal/neonatal infections greater than 80%. In infections during pregnancy, the mother usually survives. Reports of successful treatment with parenteral penicillin or ampicillin exist. Trimethoprim-sulfamethoxazole has been shown effective in patients allergic to penicillin.

A bacteriophage, *Listeria phage P100*, has been proposed as food additive to control *L. monocytogenes*. Bacteriophage treatments have been developed by several companies. EBI Food Safety and Intralytix both have products suitable for treatment of the bacterium. The U.S. Food and Drug Administration (FDA) approved a cocktail of six bacteriophages from Intralytix, and a one-type phage product from EBI Food Safety designed to kill *L. monocytogenes*. Uses would potentially include spraying it on fruits and ready-to-eat meat such as sliced ham and turkey.

Use as a Transfection Vector

Because *L. monocytogenes* is an intracellular bacterium, some studies have used this bacterium as a vector to deliver genes *in vitro*. Current transfection efficiency remains poor. One example of the successful use of *L. monocytogenes* in *in vitro* transfer technologies is in the delivery of gene therapies for cystic fibrosis cases.

Cancer Treatment

Listeria monocytogenes is being investigated as a cancer immunotherapy for several types of cancer.

A live attenuated *Listeria monocytogenes* cancer vaccine, ADXS11-001, is under development as a possible treatment for cervical carcinoma.

Epidemiology

Researchers have found *Listeria monocytogenes* in at least 37 mammalian species, both domesticated and feral, as well as in at least 17 species of birds and possibly in some species of fish and shellfish. Laboratories can isolate *Listeria monocytogenes* from soil, silage, and other environmental sources. *Listeria monocytogenes* is quite hardy and resists the deleterious effects of freezing, drying, and heat remarkably well for a bacterium that does not form spores. Most *Listeria monocytogenes* strains are pathogenic to some degree.

Routes of Infection

Listeria monocytogenes has been associated with such foods as raw milk, pasteurized fluid milk, cheeses (particularly soft-ripened varieties), ice cream, raw vegetables, fermented raw-meat sausages, raw and cooked poultry, raw meats (of all types), and raw and smoked fish. Most bacteria can survive near freezing temperatures, but cannot absorb nutrients, grow or replicate. *L. monocytogenes* ability to grow at temperatures as low as 0 °C permits exponential multiplication in refrigerated foods. At refrigeration temperature, such as 4 °C, the amount of ferric iron can affect the growth of *L. monocytogenes.*

Infectious Cycle

The primary site of infection is the intestinal epithelium, where the bacteria invade nonphagocytic cells via the "zipper" mechanism. Uptake is stimulated by the binding of listerial internalins (Inl) to E-cadherin, a host cell adhesion factor, or Met (c-Met), hepatocyte growth factor. This binding activates certain Rho-GTPases, which subsequently bind and stabilize Wiskott Aldrich syndrome protein (WAsp). WAsp can then bind the Arp2/3 complex and serve as an actin nucleation point. Subsequent actin polymerization creates a "phagocytic cup", an actin-based structure normally formed around foreign materials by phagocytes prior to endocytosis. The net effect of internalin binding is to exploit the junction-forming apparatus of the host into internalizing the bacterium. *L. monocytogenes* can also invade phagocytic cells (e.g., macrophages), but requires only internalins for invasion of nonphagocytic cells.

Following internalization, the bacterium must escape from the vacuole/phagosome before fusion with a lysosome can occur. Three main virulence factors that allow the bacterium to escape are listeriolysin O (LLO-encoded by *hly*) phospholipase A (encoded by *plcA*) and phospholipase B (*plcB*). Secretion of LLO and PlcA disrupts the vacuolar membrane and allows the bacterium to escape into the cytoplasm, where it may proliferate.

Once in the cytoplasm, *L. monocytogenes* exploits host actin for the second time. ActA proteins associated with the old bacterial cell pole (being a bacillus, *L. monocytogenes* septates in the middle of the cell, thus has one new pole and one old pole) are capable of binding the Arp2/3 complex, thereby inducing actin nucleation at a specific area of the bacterial cell surface. Actin polymerization then propels the bacterium unidirectionally into the host cell membrane. The protrusion formed may then be internalized by a neighboring cell, forming a double-membrane vacuole from which the bacterium must escape using LLO and PlcB. This mode of direct cell-to-cell spread involves a cellular mechanism known as paracytophagy.

Escherichia Coli

Escherichia coli (E. coli) bacteria normally live in the intestines of healthy people and animals. Most varieties of E. coli are harmless or cause relatively brief diarrhea. But a few particularly nasty strains, such as E. coli O157:H7, can cause severe abdominal cramps, bloody diarrhea and vomiting.

You may be exposed to E. coli from contaminated water or food — especially raw vegetables and undercooked ground beef. Healthy adults usually recover from infection with E. coli O157:H7 within a week, but young children and older adults have a greater risk of developing a life-threatening form of kidney failure called hemolytic uremic syndrome.

Symptoms

Signs and symptoms of E. coli O157:H7 infection typically begin three or four days after exposure to the bacteria, though you may become ill as soon as one day after to more than a week later. Signs and symptoms include:

- Diarrhea, which may range from mild and watery to severe and bloody.
- Abdominal cramping, pain or tenderness.
- Nausea and vomiting, in some people.

Causes

Among the many strains of E. coli, only a few trigger diarrhea. One group of E. coli — which includes O157:H7 — produces a powerful toxin that damages the lining of the small intestine, which can cause bloody diarrhea. You develop an E. coli infection when you ingest this strain of bacteria.

Unlike many other disease-causing bacteria, E. coli can cause an infection even if you ingest only small amounts. Because of this, you can be sickened by E. coli from eating a slightly undercooked hamburger or from swallowing a mouthful of contaminated pool water.

Potential sources of exposure include contaminated food or water and person-to-person contact.

Contaminated Food

The most common way to acquire an E. coli infection is by eating contaminated food, such as:

- Ground beef. When cattle are slaughtered and processed, E. coli bacteria in their intestines can get on the meat. Ground beef combines meat from many different animals, increasing the risk of contamination.

- Unpasteurized milk. E. coli bacteria on a cow's udder or on milking equipment can get into raw milk.
- Fresh produce. Runoff from cattle farms can contaminate fields where fresh produce is grown. Certain vegetables, such as spinach and lettuce, are particularly vulnerable to this type of contamination.

Contaminated Water

Human and animal feces may pollute ground and surface water, including streams, rivers, lakes and water used to irrigate crops. Although public water systems use chlorine, ultraviolet light or ozone to kill E. coli, some outbreaks have been linked to contaminated municipal water supplies.

Private wells are a greater cause for concern because they don't often have any disinfecting system. Rural water supplies are the most likely to be contaminated. Some people also have been infected after swimming in pools or lakes contaminated with feces.

Personal Contact

E. coli bacteria can easily travel from person to person, especially when infected adults and children don't wash their hands properly. Family members of young children with E. coli infection are especially likely to acquire it themselves. Outbreaks have also occurred among children visiting petting zoos and in animal barns at county fairs.

Risk Factors

E. coli can affect anyone who is exposed to the bacteria. But some people are more likely to develop problems than are others. Risk factors include:

- Age. Young children and older adults are at higher risk of experiencing illness caused by E. coli and more-serious complications from the infection.
- Weakened immune systems. People who have weakened immune systems — from AIDS or drugs to treat cancer or prevent the rejection of organ transplants — are more likely to become ill from ingesting E. coli.
- Eating certain types of food. Riskier foods include undercooked hamburger; unpasteurized milk, apple juice or cider; and soft cheeses made from raw milk.
- Time of year. Though it's not clear why, the majority of E. coli infections in the U.S. occur from June through September.
- Decreased stomach acid levels. Stomach acid offers some protection against

E. coli. If you take medications to reduce your levels of stomach acid, such as esomeprazole (Nexium), pantoprazole (Protonix), lansoprazole (Prevacid) and omeprazole (Prilosec), you may increase your risk of an E. coli infection.

Complications

Most healthy adults recover from E. coli illness within a week. Some people — particularly young children and older adults — may develop a life-threatening form of kidney failure called hemolytic uremic syndrome.

Prevention

No vaccine or medication can protect you from E. coli-based illness, though researchers are investigating potential vaccines. To reduce your chance of being exposed to E. coli, avoid risky foods and watch out for cross-contamination.

Risky foods

- Cook hamburgers until they're 160 F. Hamburgers should be well-done, with no pink showing anywhere in the meat. But color isn't a reliable indicator of whether or not meat is done cooking. Meat — especially if grilled — can brown before it's completely cooked.

That's why it's important to use a meat thermometer to ensure that meat is heated to at least 160 F (71 C) at its thickest point.

- Drink pasteurized milk, juice and cider. Any boxed or bottled juice kept at room temperature is likely to be pasteurized, even if the label doesn't say so.
- Wash raw produce thoroughly. Washing produce won't necessarily get rid of all E. coli — especially in leafy greens, which provide many spots for the bacteria to attach themselves to. Careful rinsing can remove dirt and reduce the amount of bacteria that may be clinging to the produce.

Avoid cross-contamination

- Wash utensils: Use hot soapy water on knives, countertops and cutting boards before and after they come into contact with fresh produce or raw meat.
- Keep raw foods separate: This includes using separate cutting boards for raw meat and foods, such as vegetables and fruits. Never put cooked hamburgers on the same plate you used for raw patties.
- Wash your hands: Wash your hands after preparing or eating food, using the toilet, or changing diapers. Make sure that children also wash their hands before eating, after using the bathroom and after contact with animals.

References

- Mikkola, Raimo. Food and Indoor Air Isolated Bacillus Non-Protein Toxins: Structures, Physico-Chemical Properties and Mechanisms of Effects on Eukaryotic Cells (PDF). p. 12. Retrieved 06, March 2020
- Gillespie SH, Hawkey PM (2006). Principles and practice of clinical bacteriology (2nd ed.). Hoboken, NJ: John Wiley & Sons. ISBN 9780470017968
- "Adult Health Advisor 2005.4: Bacteria in Urine, No Symptoms (Asymptomatic Bacteriuria)". Archived from the original on 2007-07-12. Retrieved 19, May 2020
- Epps SV, Harvey RB, Hume ME, Phillips TD, Anderson RC, Nisbet DJ (2013). "Foodborne Campylobacter: infections, metabolism, pathogenesis and reservoirs". International Journal of Environmental Research and Public Health. 10 (12): 6292–304. doi:10.3390/ijerph10126292. PMC 3881114. PMID 24287853
- Ryan, Kenneth J.; Ray, C. George; Ahmad, Nafees; Drew, W. Lawrence; Lagunoff, Michael; Pottinger, Paul; Reller, L. Barth; Sterling, Charles R. (2014). "Pathogenesis of Bacterial Infections". Sherris Medical Microbiology (6th ed.). New York: McGraw Hill Education. pp. 391–406. ISBN 978-0-07-181826-1
- McFall-Ngai, Margaret (2007-01-11). "Adaptive Immunity: Care for the community". Nature. 445 (7124): 153–153. doi:10.1038/445153a. ISSN 0028-0836

Permissions

All chapters in this book are published with permission under the Creative Commons Attribution Share Alike License or equivalent. Every chapter published in this book has been scrutinized by our experts. Their significance has been extensively debated. The topics covered herein carry significant information for a comprehensive understanding. They may even be implemented as practical applications or may be referred to as a beginning point for further studies.

We would like to thank the editorial team for lending their expertise to make the book truly unique. They have played a crucial role in the development of this book. Without their invaluable contributions this book wouldn't have been possible. They have made vital efforts to compile up to date information on the varied aspects of this subject to make this book a valuable addition to the collection of many professionals and students.

This book was conceptualized with the vision of imparting up-to-date and integrated information in this field. To ensure the same, a matchless editorial board was set up. Every individual on the board went through rigorous rounds of assessment to prove their worth. After which they invested a large part of their time researching and compiling the most relevant data for our readers.

The editorial board has been involved in producing this book since its inception. They have spent rigorous hours researching and exploring the diverse topics which have resulted in the successful publishing of this book. They have passed on their knowledge of decades through this book. To expedite this challenging task, the publisher supported the team at every step. A small team of assistant editors was also appointed to further simplify the editing procedure and attain best results for the readers.

Apart from the editorial board, the designing team has also invested a significant amount of their time in understanding the subject and creating the most relevant covers. They scrutinized every image to scout for the most suitable representation of the subject and create an appropriate cover for the book.

The publishing team has been an ardent support to the editorial, designing and production team. Their endless efforts to recruit the best for this project, has resulted in the accomplishment of this book. They are a veteran in the field of academics and their pool of knowledge is as vast as their experience in printing. Their expertise and guidance has proved useful at every step. Their uncompromising quality standards have made this book an exceptional effort. Their encouragement from time to time has been an inspiration for everyone.

The publisher and the editorial board hope that this book will prove to be a valuable piece of knowledge for students, practitioners and scholars across the globe.

Index

www.ingramcontent.com/pod-product-compliance
Lightning Source LLC
LaVergne TN
LVHW080730210926
848485LV00003B/544
* 9 7 8 1 6 3 9 8 7 3 7 2 2 *